Stochastic Space–Time Models and Limit Theorems

Mathematics and Its Applications

V. 19

Managing Editor:

M. HAZEWINKEL
Centre for Mathematics and Computer Science, Amsterdam, The Netherlands

Editorial Board:

R. W. BROCKETT, *Harvard University, Cambridge, Mass., U.S.A.*
J. CORONES, *Iowa State University, U.S.A. and Ames Laboratory, U.S. Department of Energy, Iowa, U.S.A.*
F. CALOGERO, *Universita degli Studi di Roma, Italy*
Yu. I. MANIN, *Steklov Institute of Mathematics, Moscow, U.S.S.R.*
A. H. G. RINNOOY KAN, *Erasmus University, Rotterdam, The Netherlands*
G.-C. ROTA, *M.I.T., Cambridge, Mass., U.S.A.*

Stochastic Space–Time Models and Limit Theorems

edited by

L. Arnold

and

P. Kotelenez

Forschungsschwerpunkt Dynamische Systeme,
Universität Bremen, F.R.G.

D. Reidel Publishing Company

A MEMBER OF THE KLUWER ACADEMIC PUBLISHERS GROUP

Dordrecht / Boston / Lancaster

Library of Congress Cataloging in Publication Data

CIP

Main entry under title:

Stochastic space-time models and limit theorems.

(Mathematics and its applications)
Includes index.
1. Stochastic analysis–Addresses, essays, lectures. 2. Stochastic differential equations–Addresses, essays, lectures. 3. Limit theorems (Probability theory)–Addresses, essays, lectures. 4. State-space methods–Addresses, essays, lectures. I. Arnold, L. (Ludwig), 1937– II. Kotelenez, P. (Peter), 1943– . III. Series: Mathematics and its applications (D. Reidel Publishing Company)
QA274.2.S776 1985 519.2 85-10895
ISBN 90-277-2038-X

Published by D. Reidel Publishing Company
P.O. Box 17, 3300 AA Dordrecht, Holland

Sold and distributed in the U.S.A. and Canada
by Kluwer Academic Publishers,
190 Old Derby Street, Hingham, MA 02043, U.S.A.

In all other countries, sold and distributed
by Kluwer Academic Publishers Group,
P.O. Box 322, 3300 AH Dordrecht, Holland

Printed in The Netherlands

Table of Contents

Part II. Limit Theorems

SERIES EDITOR'S PREFACE

Approach your problems from the right end and begin with the answers. Then one day, perhaps you will find the final question.

'The Hermit Clad in Crane Feathers' in R. van Gulik's The Chinese Maze Murders.

It isn't that they can't see the solution.
It is that they can't see the problem.

G.K. Chesterton. The Scandal of Father Brown 'The Point of a Pin'.

Growing specialisation and diversification have brought a host of monographs and textbooks on increasingly specialized topics. However, the "tree" of knowledge of mathematics and related fields does not grow only by putting forth new branches. It also happens, quite often in fact, that branches wich were thought to be completely disparate are suddenly seen to be related.

Further, the kind and level of sophistication of mathematics applied in various sciences has changed drastically in recent years: measure theory is used (non-trivially) in regional and theoretical economics; algebraic geometry interacts with physics; the Minkowsky lemma, coding theory and the structure of water meet one another in packing and covering theory; quantum fields, crystal defects and mathematical programming profit from homotopy theory; Lie algebras are relevant to filtering; and prediction and electrical engineering can use Stein spaces. And in addition to this there are such new emerging subdisciplines as "experimental mathematics", "CFD", "completely integrable systems", "chaos, synergetics and large-scale order", which are almost impossible to fit into the existing classification schemes. They draw upon widely different sections of mathematics. This programme, Mathematics and Its Applications, is devoted to new emerging (sub)disciplines and to such (new) interrelations as exampla gratia:

-a central concept which plays an important role in several different mathematical and/or scientific specialized areas;
-new applications of the results and ideas from one area of scientific endeavour into another;
-influences which the results, problems and concepts of one field of enquiry have and have had on the development of another.

The Mathematics and Its Applications programme tries to make available a careful selection of books which fit the philosophy outlined above. With such books, which are stimulating rather than definitive, intriguing rather than encyclopaedic, we hope to contribute something towards better communication among the practitioners in diversified fields.

It's remarkable how fast sometimes a new coherent mathematical subdiscipline can establish itself. One such area is concerned with stochastic space-time models including pattern formations and dynamically evolving patterns in space. This is a topic, which as a recognized entity, did, perhaps even less than a decade ago, not yet exist, and now the small part of it which goes under the name of reaction-diffusion equations has already at least two books published and another in preparation.

As usual, these new fields tend not to fit very well in existing classification schemes, making than a prime area of concern of this book series.

It's also interesting to see how the evolution mathematical thinking (and applications) repeats itself (more or less) in patterns. In deterministic mathematics there are the lines of development scalars → vectors → infinite dimensional spaces and static → dynamic. And so, after stochastic variables, stochastic processes, and stochastic differential equations have established themselves as substantial fields of investigation, the time seems to have come for stochastic operators, stochastic partial differential equations and stochastic functional analysis.

There is much more involved than just generalizing everything in sight to a stochastic context. There are deep fundamental interactions between geometry (of the underlying spaces) and stochastics on the one hand and stochastics and dynamics - the main concern of the present volume - on the other. There are lots of surprises, especially for such as myself who were mainly trained deterministically.

That is what makes - in my view - the field so fascinating, as the thirteen state-of-the-art surveys in this volume will testify.

The unreasonable effectiveness of mathematics in science....

Eugene Wigner

Well, if you know of a better 'ole, go to it.

Bruce Bairnsfather

What is now proved was once only imagined.

William Blake

As long as algebra and geology proceeded along separate paths, their advance was slow and their applications limited.
But when these sciences joined company they drew from each other fresh vitality and thenceforward marched on at a rapid pace towards perfection.

Joseph Louis Lagrange

Bussum, March 1985

Michiel Hazewinkel

PREFACE

Stochastic space-time models describe phenomena which change with time, are distributed in space and contain fluctuations. Those models are of growing importance in various fields such as physics and chemistry (many particle systems, quantum field theory, reaction-diffusion models), biology (population dynamics) and engineering (random loads on mechanical structures). They also give rise to challenging mathematical problems. The mathematics of space-time phenomena, which is just about to emerge, is basically a combination of the modern theory of stochastic processes and functional analysis.

The present volume presents the state of the art in this new field. It contains 13 invited papers of a workshop held in November 1983 at the University of Bremen delivered by leading specialists in the field. The papers are preceded by a unifying introduction. The main subjects are:

(i) Stochastic partial differential equations (existence, uniqueness and regularity of solutions),

(ii) Stochastic analysis and Markov processes in infinite dimensions,

(iii) Limit theorems, applications.

Finally, we want to thank the Reidel Publishing Company for their friendly cooperation and the University of Bremen for the financial support to our workshop.

P. Kotelenez

STOCHASTIC SPACE-TIME MODELS AND LIMIT THEOREMS: AN INTRODUCTION

1. Stochastic Partial Differential Equations (SPDE's) as Stochastic Space-Time Models

Stochastic space-time models describe phenomena which change with time, are inhomogeneous in space and depend on chance. They arise in various fields of physics, chemistry, biology and engineering. Both analogy to the theory of spatially homogeneous stochastic systems and other reasoning suggest that a class of stochastic space-time phenomena could be modelled as solutions of SPDE's which can be formally written as a (deterministic) partial differential equation (PDE) perturbed by a "suitable" noise term

$$\frac{\partial}{\partial t}X(t,r,\omega) = P(t,r,X,\partial_i X) + \xi(t,r,X,\partial_i X,\omega) \tag{1}$$

(initial condition, boundary condition - if necessary).

Here $r \in D$, where D is some open domain in R^n and P is a PD operator, densely defined on $L_2(D)$.

There are essentially two sources for the stochasticity in (1), namely internal noise and external noise.

In the case of internal noise the unperturbed PDE($\xi \equiv 0$ in (1)) is the phenomenological or macroscopic description of a system which consists of a large number of subsystems, like the (average) density of a chemical reactant in a reactor. Thus (1) with $\xi \equiv 0$ "is only approximate, and in reality there will be small deviations from them which show up as fluctuations" (van Kampen [8]). These fluctuations are modelled by $\xi(t,r,X,\partial_i X,\omega)$, and (1) describes a special class of synergetic

L. Arnold and P. Kotelenez (eds.), Stochastic Space-Time Models and Limit Theorems, 1–9.

systems in the sense of Haken [4] (cf. Section 3 for a more detailed example as well as Nicolis and Prigogine [13]).

External noise appears by exposing a deterministic system to a random force. In engineering many phenomena are of this type (cf. Krée for a general analysis of such systems and Kotelenez, Section II, Ex. 3, for an example from continuum mechanics).

2. On the Solvability of SPDE's

By analogy to the spatially homogeneous (finite dimensional) case one sets

$$dM(t,\cdot,X,\partial_i X,\omega) = \xi(t,\cdot,X,\partial_i X,\omega)dt$$

and interprets (1) as an infinite dimensional Itô (evolution) equation with unbounded "drift" operator P . Then one tries to apply one of the standard PDE methods to (1), e.g., "variation of constants" if P is linear and its closure generates a strongly continuous semigroup, or the variational approach if P is nonlinear and satisfies the variational assumptions. In the case of a second order equation one transforms this equation into a 2-system of first order equations on a suitable product space and obtains an evolution equation of type (1) with unbounded drift.

Now let $\xi(t,r,X,\partial_i X,\omega) \equiv \xi(t,r,\omega)$ be the standard white noise in t and r . Then $M(t)$ is the cylindrical Brownian motion on $L_2(D)$, which defines for each t only a weak (finitely additive) Gaussian distribution on the cylinder sets of $L_2(D)$ with characteristic functional $\exp(-\frac{t}{2}|\phi|^2)$, where $|\cdot|$ is the standard Hilbert norm of $L_2(D)$ and $\phi \in L_2(D)$ (cf. Kuo [11]). However, since $M(t)$ is not σ-additive on $L_2(D)$ we cannot treat (1) as an Itô evolution equation on $L_2(D)$. On the other hand, we know from the theory of abstract Wiener spaces that $M(t)$ defines a σ-additive measure on some enlarged Banach (distribution) space $B \supset L_2(D)$ (Kuo [11], Itô [6]). Consequently, one could try to extend P to an unbounded operator $\tilde{P}$, which satisfies on B the deterministic PDE assumptions, in order to give a meaning to (1) as an SPDE

(Itô evolution equation) on B. It is easy to see that this is even for linear operators in general not possible (sufficient conditions for extendibility of linear operators to certain distribution spaces are given in DaPrato and Grisvard [2]). Moreover, since there is no general extension of the pointwise multiplication of continuous functions to the distributions (Schwartz [14]), we cannot extend a nonlinear P with polynomial part (cf. the example of Section 3). But there may be a weak dimension dependent solvability of (1) for certain nonlinear P on bounded domains D and cylindrical Brownian motion $M(t)$ as an equation on $L_2(D)$ (without extending!), using the smoothening property of the semigroup generated by the (linear) main part of P (cf. Section 4 and Kotelenez). On the other hand, Dawson [3] has obtained a diffusion approximation to a system of branching Brownian motions which is measure valued and satisfies an equation of type (1) with state dependent noise (driven by the cylindrical Brownian motion) only formally, i.e., the operations in (1) are not separately defined. In short, if $M(t)$ does not define a σ-additive measure on $L_2(D)$, the treatment of (1) as an Itô evolution equation (in the sense described above) excludes interesting stochastic space-time models which formally satisfy (1). Therefore, one may wonder whether we cannot drop the axiom of σ-additivity (Kolmogorov [9]) and solve (1) directly on $L_2(D)$ in the setting of weak (finitely additive) measures (cf. Kallianpur and Karandikar [7] for applications of weak measures in nonlinear filtering). To our knowledge this is still an open problem.

3. A Key Example

Let us now discuss an example, where (1) arises in a "natural" way from the unperturbed PDE. For the sake of ease we assume D to be a bounded convex domain with smooth boundary. Set

$$P(t,r,X,\partial_i X) = \Delta X + f(X), \tag{2}$$

where Δ is the Laplacian closed w.r.t. homogeneous

Dirichlet boundary conditions and f is a polynomial with constant coefficients s.t. $f(o) \geq 0$ and for nonlinear f the leading coefficient is negative. Moreover, we assume a positive initial condition $X(o,r) \geq 0$. Then, for $\xi \equiv 0$, (1) becomes the classical reaction and diffusion equation for the density of one reactant in the reactor D, which has a positive solution $X = X(t,r)$. The validity of this (macroscopic) description rests on the implicit assumption that the number of particles in D is large (~infinite) - as indicated in Section 1. Consequently, we can try to derive X as the thermodynamic limit of an appropriately rescaled sequence of stochastic (counting) processes for the density of the reactant, if the number of particles tends to infinity. This can be done as follows: Divide D into N cells of volume v (at least in the interior) and define a jump Markov process $X^{v,N}$ by counting the number of particles in each cell and dividing this number by v, where discretizations of Δ and $f(x)$ are used to describe the random change in the number of particles due to diffusion (between cells) and reaction (within cells). Arnold and Theodosopulu [1] have shown that for a certain sequence (v,N) s.t. $vN \to \infty$, $N \to \infty$

$$X^{v,N} \longrightarrow X , \tag{3}$$

which is called the law of large numbers (LLN). The next canonical step is to compute the fluctuations around X by a central limit theorem (CLT), i.e., find a sequence of positive numbers $\alpha(v,N)$ s.t.

$$\alpha(v,N)(X^{v,N}-X) \Rightarrow Y , \tag{4}$$

where Y is some Ornstein-Uhlenbeck process. Y is called the correction term to the deterministic PDE for the density X, i.e., if (4) is correct, we have $X^{v,N} \sim \frac{1}{\alpha(v,N)} Y$.

The third problem is to describe Y, if possible, as the solution of a (linear) SPDE

$$dY(t) = AY(t)dt + dM(t), \tag{5}$$

where $M(t)$ is some Gaussian martingale and A is an unbounded "drift". If, e.g. the ratio $\frac{v}{N}$ is

chosen in such a way that $M(t)$ is white noise in the spatial direction or of the order of the derivative of white noise then it is desirable to obtain spatially smooth versions of Y.
The CLT (4) (and the LLN (3)) was obtained by Kotelenez [10] under the assumption that the reaction is linear, and it was shown that Y satisfies (5) with $M(t)$ of the order of the derivative of white noise in the spatial direction and $A = \tilde{\Delta} + f'$, where $\tilde{\Delta}$ is the extension of Δ to appropriately chosen Hilbert distribution spaces. The mathematical difficulty for getting a Gaussian distribution valued Y as correction term in the case of nonlinear reaction (which is an example of local interaction of particles) is linked to the problem of Section 2, namely that there is no general extension of the pointwise multiplication to the distributions. Moreover, "correcting" the nonlinear PDE for the density X by white noise seems to contradict physical observation. Consequently, physicists suggest a different scaling for nonlinear reaction and diffusion systems near bifurcation points (Van den Broeck).

4. Contents of these Proceedings

The following problems emerge from the previous considerations:
a) Describe stochastic space-time models as (possibly distribution valued) stochastic processes Y.
b) Derive Y as the limit of stochastic processes on lattices or of positions of finitely many particles etc., where we can easily define interaction through multiplication and other mathematical operations. These problems were the major topics of our workshop.

To a) Typically Y will live on a nuclear space Φ' which is the strong dual of a nuclear space Φ (of smooth elements) with a Hilbert space H (e.g. $H = L_2(D)$) in between, where H is identified with its dual H',

$$\Phi \subset H = H' \subset \Phi' . \qquad (6)$$

If Φ is also a Fréchet space then it is the projective limit of a sequence of Hilbert spaces H_α, whence Y can be analyzed on a scale of Hilbert spaces

$$\Phi \subset H_\alpha \subset H_0 = H_0' \subset H_{-\alpha} \subset \Phi' \, , \tag{7}$$

where $\alpha \in N$ (or R_+), $H_0 := H$ and $H_{-\alpha} = H_\alpha'$ with dense continuous imbeddings and the nuclearity property. It has been shown in Section II of Kotelenez that (1) makes sense on some $H_{-\alpha}$ for $H_{-\alpha}$-valued state independent noise $M(t)$ and linear P by extending P. Moreover, if $M(t)$ is state dependent, (1) can be given a rigorous meaning if the main part of P is linear and generates an analytic semigroup $U(t)$ s.t. $U(t)M(t) \in H_0$ (cf. Kotelenez). Thus, at least a certain class of space-time models with distribution valued noise input can be represented by SPDE's (interpreted as Itô evolution equations) on some Hilbert space. The papers of Ichikawa, DaPrato and Section I of Kotelenez are concerned with properties of SPDE's on a fixed abstract Hilbert space.

As explained in Section 2 the treatment of (1) depends on the Itô integral in infinite dimensions. This integral has been well described by Metivier and Pellaumail [12] if the state space is a Hilbert space. On the other hand, even if we can show existence and uniqueness of (1) (as well as tightness of some approximation to (1)) on some Hilbert space, this space may be not the most "natural" space. An example is the Ornstein-Uhlenbeck process Y obtained by Holley and Stroock [5] as the Gaussian approximation to a system of branching Brownian motions, which solves (5) on some Hilbert distribution space but is ergodic on some Banach distribution space. Thus, it is desirable to extend the Itô integral to Banach spaces, which is done by Dettweiler. Ustunel deals with generalizations of basic stochastic concept to the nuclear triple (6), including SPDE's with a finite dimensional Wiener process as driving force. However, we have seen in Section 2 that the concept of an SPDE of type (1) may be too narrow for many nonlinear models. A different approach to nonlinear stochastic space-time models (linked to quantum field theory) is given by Albeverio, Høegh-Krohn and Holden. Sections I and II of their paper deal with the construction of Markov fields through Dirichlet forms on infinite dimensional spaces, Section III with the construction of Markov surfaces, i.e., of random fields which do not depend on points but on (n-1)-dimensional hypersurfaces and take values in a Lie group. Krée's paper deals with the description of SPDE's arising from external noise. The (generalized stationary) Gaussian driving term is obtained as the solution to the linear SPDE (5), where A generates an asymptotically stable linear semigroup. Moreover, smoothness properties of probability measures on infinite dimensional spaces are derived in the framework of Banach valued Sobolev spaces.

To b) In Rost's paper weak convergence of a system of interacting particles to a generalized Ornstein-Uhlenbeck process is proved in the set-up (7). Zessin gives a weak convergence criterion in the set-up (6) in terms of moments. Grigelionis and Mikulevičius derive weak convergence criteria for semimartingales with values in a rigged Hilbert space

$$B \subset H = H' \subset B' \tag{8}$$

by extending finite dimensional results in terms of local characteristics to (8), where H is a Hilbert space identified with its dual, B is a Banach space densely and continuously imbedded into H and B' its dual. In Metivier a weak compactness criterion for Hilbert space valued martingales is given and Pardoux proves a wide-band noise approximation for a semi-linear SPDE. Finally, Van den Broeck deals with the problem of Gaussian and non-Gaussian approximation for nonlinear particle systems (cf. Section 3).

Peter Kotelenez
Forschungsschwerpunkt Dynamische Systeme
Universität Bremen
Bibliothekstraße
Postfach 330 440
2800 Bremen 33
West Germany

5. References

[1] L. ARNOLD and M. THEODOSOPULU: "Deterministic Limit of the Stochastic Model of Chemical Reactions with Diffusion", Adv. Appl. Prob. 12 (1980) 367 - 379

[2] G. DA PRATO and P. GRISVARD: "Maximal Regularity for Evolution Equations by Interpolation and Extrapolation" to appear in J. functional Analysis

[3] D. A. DAWSON: "Stochastic Evolution Equations and Related Measure Processes", J. Multivariate Anal. 5 (1975) 1 - 52

[4] H. HAKEN: "Advanced Synergetics", Springer-Verlag, Berlin-New York 1983

[5] R. HOLLEY and D. W. STROOCK: "Generalized Ornstein-Uhlenbeck Processes and Infinite Particle Branching Brownian Motions", Publ. RIMS, Kyoto Univ. 14 (1978), 741 - 788

[6] K. ITÔ: "Continuous Additive S'-Processes", in B. Grigelionis (ed.) "Stochastic Differential Systems", Springer Verlag, Berlin-New York 1980

[7] G. KALLIANPUR, R. L. KARANDIKAR: "White Noise Calculus and Nonlinear Filtering Theory" to appear in Annals of Probab.

[8] N. G. VAN KAMPEN: "Stochastic Processes in Physics and Chemistry", North Holland, Amsterdam-New York 1983

[9] A. N. KOLMOGOROV: "Grundbegriffe der Wahrscheinlichkeitsrechnung" Ergebn. d. Math. 2, Heft 3, Berlin 1933

[10] P. KOTELENEZ: "Law of Large Numbers and Central Limit Theorem for Chemical Reactions with Diffusion", Ph. D. Thesis, Bremen 1982

[11] H. H. KUO: "Gaussian Measures in Banach Spaces", Springer Verlag, Berlin-New York 1975

[12] M. METIVIER and J. PELLAUMAIL: "Stochastic Integration", Academic Press, New York-London 1980

[13] G. NICOLIS and I. PRIGOGINE: "Self-Organization in Non-equilibrium Systems", John Wiley & Sons, New York-London 1977

[14] L. SCHWARTZ: "Sur l'impossibilité de la multiplication des distribution", C. R. Acad. Sci., Paris 239 (1954), 847 - 848

S.Albeverio, R.Høegh-Krohn, H.Holden

MARKOV PROCESSES ON INFINITE DIMENSIONAL SPACES, MARKOV FIELDS AND MARKOV COSURFACES

ABSTRACT: We review work on Dirichlet forms and symmetric Markov processes on infinite dimensional spaces. Especially we consider the connections with the construction of homogeneous generalized Markov random fields. We also discuss a non commutative extension to the case where the state space is a group. The extension involves a stochastic calculus for group valued mappings defined on hypersurfaces of codimension 1.

1. INTRODUCTION

The theory of Dirichlet forms and of symmetric Markov processes was developed as a powerful L^2-extension of classical potential theory and the associated processes. The basic theory is beautifully presented in [1], [2], [3] (for newer developments see also [4]). An ideal field of application of the theory is quantum mechanics, where the L^2-extension of classical C-results is needed, both because the natural space to work with is a Hilbert space and because the drift coefficients and potential functions involved in important examples have singularities of various sorts. In particular it was soon realized that the usual existence and uniqueness results of stochastic differential equations can not be applied to the stochastic equations arising in quantum mechanical situations (via the stochastic mechanical formulation of quantum mechanics [5], [6]). A stochastic calculus based on the theory of Dirichlet forms was developed, simultaneously, for the needs of quantum mechanics, by Albeverio, Høegh-Krohn and Streit in [7], [8] and in generality by Fukushima in [1], [9], [10]. Other examples of a close interaction between quantum mechanical problems and the theory of Dirichlet forms are provided by the study of criteria for closability of Dirichlet forms [1], [8], [11], [12], [13], [14], the problem of approximation of Dirichlet forms and associated processes [13], [15],

L. Arnold and P. Kotelenez (eds.), Stochastic Space-Time Models and Limit Theorems, 11–40.

[16], [71], the problem of uniqueness of Dirichlet forms and associated self-adjoint operators [1], [7], [8], [17], [18], [19], [71], the problem of the relation between Dirichlet forms and Schrödinger operators [6-8], [16], [20-25], the problem of ergodicity and barriers for the processes [4], [5], [6], [26-32], the problem of densities for the processes [3][4]and many other.

For some applications of the theory of Dirichlet forms outside quantum mechanics see e.g. [1-3], [14], [23], [34-42]and references therein.
The theory of Dirichlet forms required for all these applications is the basic one in which the underlying space is locally compact. For applications to quantum field theory is was realized quite early [7] that an extension to the case of an underlying infinite dimensional linear space (e.g. a Hilbert space) is needed. It can actually be shown [43] that every local Dirichlet form can be looked upon as a Dirichlet form associated with some Hilbert space, so that the theory of Dirichlet forms associated with Hilbert spaces properly extends the usual one on locally compact spaces, at least in the local (diffusion) case.
The construction of a Dirichlet form and associated processes on infinite dimensional spaces was begun by Albeverio and Høegh-Krohn in [7], subsequently worked by the same authors in [44], [45] (see also [23], [46] for applications) and by Paclet [47], [48] and Kusuoka [37] (see also [49] for a recent application).
More recently Albeverio, Fenstad, Høegh-Krohn and Lindstrøm [50] have given a version of the theory of Dirichlet forms in terms of non standard analysis, which unifies the finite dimensional and the infinite dimensional theory.

In Sect. I of this paper we discuss some basic properties and some new results on Dirichlet forms, both in the finite dimensional and in the infinite dimensional case. The study of Dirichlet forms is actually equivalent with the study of symmetric Markov processes. In the case where the state space is an infinite dimensional function or distributional space it appear quite natural to try to look at the processes as random fields or generalized random fields. An especially interesting class would be the one in which the (generalized) random fields have themselves a Markov property and are homogeneous (which extends the stationarity and symmetry properties of the processes). It

was realized by Nelson, see e.g. [51], and in [7] that quantum fields provide, at least in principle, natural examples of such fields. Technically, the construction of global homogeneous Markov fields has turned out to be very difficult, but by now several examples are known [52-54] (see also [56] for expositions of the method introduced by Albeverio and Høegh-Krohn in [52]).
The construction involves, among other things, the solution of a stochastic Dirichlet problem, with distributional data, which have been recently treated in great generality with powerful methods of axiomatic potential theory [57-60] (see also [61] for related methods).
In Sect. II of this paper we shall discuss briefly these connections.
There are several directions in which extensions of the ideas of Sect. I and II can be looked for, and again the inspiration has come from problems of quantum theory (quantum gauge fields). The case where the underlying space has the richer structure of a manifold has been discussed by Albeverio and Høegh-Krohn in [23].
Extensions to the cases where the state space is a group have been obtained following two methods. One inspired by the theory of group representations, providing algebraic analogues of the random fields. This has lead to a series of papers [62], [63](and references therein), which have been partly reviewed elsewhere [23], so that this method will not be discussed here.
Another method, which is presented here, is the one in which the pointwise defined random field is replayed by a random field associated with d-1-dimensional hypersurfaces in a d-dimensional manifold, the value being again in a group. This approach leads in particular to the development of a theory of stochastic integrals for d-1-forms which extends the theory of Brownian motion on a Lie group, and of the associated integrals, to the case of multidimensional time parameters.
Sect. III in the present paper reviews some of the recent developments in this area.

I. Dirichlet forms and symmetric Markov processes on infinite dimensional spaces

In quantum field theory the energy operator (Hamiltonian) is formally given by an analogue of the classical Dirichlet forms, namely by

$$\frac{1}{2} \int (\nabla f)^2 d\nu ,$$

with ν a certain measure on an infinite dimensional linear space (the so called "time zero vacuum measure") and f any smooth cylinder function, ∇f being then a naturally defined gradient of f. The mathematical realization of this idea has been given by Albeverio and Høegh-Krohn in [44] (see also [7], [45]). The study of Dirichlet forms of the above type (and of related ones) has been undertaken also from other points of view by S. Kusuoka [37] and Ph. Paclet [47], [48], see also [50] for a recent formulation in terms of non-standard analysis.
Let us here shortly illustrate the main ideas of the approach [44], recalling at the same time the corresponding problems in finite dimensions.

In the finite dimensional situation ν is a Radon measure with support $\mathbb{R}^d$ (but the general theory holds also for locally compact Hausdorff space, with countable base for the topology). A Dirichlet form E on $L^2(\nu)$ is then a bilinear symmetric, nonnegative, closed form, with dense domain $D(E) \subset L^2(\nu)$ s.t. $E(f^{\#},f^{\#}) \leq E(f,f)$ for $f^{\#} = (f \vee 0) \wedge 1$, for all $f \in D(E)$ (the fact that E "contracts" under $f \to f^{\#}$ is the "Dirichlet property"). The importance of such forms is that they are in 1 - 1 correspondence with (ν-) symmetric Markov semigroups $P_t = e^{-tH}$, $t \geq 0$, in $L^2(\nu)$, with H self-adjoint, nonnegative, where Markov means $0 \leq f \leq 1 \to 0 \leq P_t f \leq 1$, $\forall f \in L^2(\nu)$ and ν-symmetric means that the adjoint P_t^* of P_t in $L^2(\nu)$ coincides with P_t itself. The correspondence $E \leftrightarrow (P_t,\ t \geq 0)$ is given by $E(f,f) = (H^{1/2}f, H^{1/2}f)_{L^2(\nu)}$, with $(\ ,\ )_{L^2(\nu)}$ the scalar product in $L^2(\nu)$. Symmetric Markov semigroups give rise to symmetric Markov processes X_t^ν with

invariant measure ν.
If E is a regular Dirichlet form then to it there is naturally associated a Hunt process. This process is a diffusion process (in the sense of continuous paths) iff the regular Dirichlet form is local.
Of course there are subleties in this general results, but we can here only refer to the literature for more details.

In quantum mechanics natural Dirichlet forms are given first on $C_0^\infty(\mathbb{R}^d)$ by $E_0(f,f) = \frac{1}{2} \int_{\mathbb{R}^d} (\nabla f)^2 d\nu$, with $\nabla = (\frac{\partial}{\partial x^1} , \ldots, \frac{\partial}{\partial x^d})$ the gradient operator.

1) For which Radon measures ν does E_0 have a closed extension? Is it then indeed a Dirichlet form? A regular one? A local one? Criteria for ν to yield closed extensions of E_0 are known in the literature, and have been reviewed in [11] (see also [14]). They are basically of 2 types, requiring ν to have a density with respect to Lebesgue measure $\lambda(dx) = dx$ (this is also necessary for d=1 [12]) and in addition either $\rho > 0$ λ-a.e., $\frac{\partial}{\partial x^i} \rho^{1/2} \in L^2_{loc}(U)$, $U \subset \mathbb{R}^d$ open, $\nu(\mathbb{R}^d - U) = 0$ or else $\rho > 0$ on compacts, $\rho \in L^1_{loc}(U)$, $U \subset \mathbb{R}^d$ open, s.t. $\lambda(\mathbb{R}^d - U) = 0$.
Regularity holds always for the closure of E_0, wherever it exists, but might fail for other closed extensions. Locality always holds for the closure (not so for other extensions [1]).

2) Is there some uniqueness, so that all closed extensions of E_0 coincide? This has been proven e.g. for $\rho > 0$, locally Lipschitz by N. Wielens [19].
One can prove that this is equivalent with the self-adjoint operator $\mathcal{L}$ associated with the closure of E_0 to be already essentially self-adjoint on $C_0^\infty(\mathbb{R}^d)$. The weaker property of "Markovian uniqueness", to the extent that all self-adjoint extensions of $\mathcal{L}$ leading to Markov semigroups (Dirichlet forms) coincide has been discussed in [17] and [19].

Two important closed extensions of E_o are the "minimal" one E, the closure of E_o, and the "maximal one" $\widetilde{E}$ with domain $D(\widetilde{E}) = \{f \in L^2(\nu) \mid \nabla f \in L^2(\nu)\}$.

3) Study of the symmetric semigroup associated with the Dirichlet form, construction of a process associated with it, construction of a "regular" realization (diffusion process), study of the corresponding potential theory. In the finite dimensional case these question have been studied to a great extent, we refer to [1-3] for basic references and [4], [7], [8], [13], [15], [16-25], [27-30], [33], [40-46], [50], [56-61], [64], [71] for some applications to quantum mechanics and other domains of physics.

4) Questions of ergodicity of P_t or of ν with respect to space translations have been discussed mathematically [6-8], [27-32], [44], [45], also for their direct interpretation in physical terms. In particular connections between capacity zero sets, nodes of ρ and tunneling - non tunneling across the nodes of ρ have been discussed, as well as connections between capacity and scattering theory for the associated Hamiltonians (cfr. 5)). Recently this has also acquired new actuality through the discussion of the method of stochastic quantization (propagated particularly by Parisi and Wu).

5) Correspondence between P_t and the "Hamiltonian semigroup" T_t in $L^2(\mathbb{R}^d, \lambda)$ given by $T_t = \varphi^{-1} P_t \varphi$ if $\rho = \varphi^2$, $\varphi > 0$ λ-a.e. this has been discussed in [7], [8][24], [25]; a very recent reference on this topics with further references is [25].

6) Approximation of P_t, the corresponding process, and of T_t in 5) by corresponding objects constructed with approximations φ_n of φ, with suitable φ_n (so that one has a good control on the corresponding P_t^n, T_t^n constructed from φ_n in the way as P_t, T_t was constructed from φ). For these problems see [15], [13] and the recent paper [16].

We shall now shortly mention few results on corresponding problems for the infinite dimensional situation.

1) Similarly as in the finite dimensional situation the starting point is the densely defined, not yet closed, form $\frac{1}{2}\int(\nabla f)^2 d\nu$ in $L^2(\nu)$. ν is here a Radon probability measure on some infinite dimensional linear space Q', realizing in some sense $\mathbb{R}^d$ for $d = \infty$. The choice we made (in [7], [44], [45]) is Q' the dual of a countably normed nuclear space Q such that $Q \subset Q'$ and such that there exists a real separable Hilbert space $\mathcal{H}$ with $Q \hookrightarrow \mathcal{H} \hookrightarrow Q'$, with continuous injections $\hookrightarrow$. $\mathcal{H}$ is used to define ∇ in as much as ∇ is defined as a map from the space FC^∞ of smooth function on Q' with base in finite dimensional subspaces of H spanned by elements in Q with values in $L^2(\nu) \times \mathcal{H}$ (in finite dimensions ∇ can correspondingly be looked upon as a map from $C^\infty(\mathbb{R}^d)$ into $L^2(\nu) \hat{\otimes} \mathbb{R}^d$). Let then $E_o(f,f) = \frac{1}{2}\int(\nabla f \cdot \nabla f)_{\mathcal{H}} d\nu$ for $f \in FC^\infty$, with $(\ ,\ )_{\mathcal{H}}$ the scalar product in $\mathcal{H}$. In [7], [40], [41] we defined as a realization of a "Dirichlet form extending E_o" as the closure of E_o in $L^2(\nu)$, whenever the closure exists. Sufficient conditions for ν such that the closure exists have been given in [45] (which extends slightly the results of [7], [44]) and correspond to conditions of the type $\frac{\partial}{\partial x^i}\rho^{1/2} \in L^2(\mathbb{R}^d)$ in finite dimensions,

namely that $1 \in D(\nabla_i^*)$ with ∇_i the directional derivative in the direction of the basis vector $e_i \in Q$ in $\mathcal{H}$, where D means domain. Other realizations of Dirichlet forms in infinite dimensions are given by Kusuoka in [37]. Here basically Q' is replaced by a separable Banach space B and Q by the dual of B. The construction of E is replaced by $\tilde{E}(f,f) = \frac{1}{2}\int(Df,Df)_{\mathcal{H}} d\nu$ with domain $D(\tilde{E}) = \{f \in L^2(\nu) \mid \int(Df,Df)_{\mathcal{H}}\, d\nu < \infty$. Df is defined for any real measurable f on B which is both "ray absolutely continuous" (in the sense that for all $q \in B$ there exists a real measurable $\tilde{f}_q$ on B s.t. $f = \tilde{f}_q$ ν-a.s. and

$f(\cdot + tq)$ is absolutely continuous in $t \in \mathbb{R}$) and "stochastic Gâteaux differentiable with respect to ν" (in the sense that $\frac{1}{t}[f(\cdot+tq)-f(\cdot)]$ converges as $t \downarrow 0$, ν-stochastically). One sets $(Df,q) = \lim_{t\downarrow 0} \frac{1}{t}[f(\cdot+tq)-f(\cdot)]$ (so that (Df,q) is the stochastic Gâteaux derivative of f in the direction q). Under point 2 we shall discuss a little the relation between $\tilde{E}$ and the above E. Yet another realization of Dirichlet forms in infinite dimensions has been given by Paclet in [47], [48]. It uses Q, Q' both separable real Hilbert space, s.t. $\mathcal{H} \subset Q'$, $Q \subset \mathcal{H}$ with Hilbert-Schmidt injections i resp. i', the transpose of i (after identification of $\mathcal{H}$ with its dual). ∇ is defined in a natural way on the space $\mathbf{H}\, C_b^1(Q')$ (of universally Lusin measurable [65], [66]) bounded functions f s.t. for all $\xi \in Q'$, $h \to f_\xi(h)$ is Fréchet C^1 on $\mathcal{H}$ and $\xi \to \nabla f_\xi(0)$ is bounded universally Lusin-measurable), with domain $D(\nabla) = \{f \in L^2(\nu) \mid \nabla f \in L^2(\nu)\}$. Both Kusuoka and Paclet thus operate rather with an infinite dimensional extension of the concept of maximal Dirichlet form $\tilde{E}$ in finite dimension (cfr. point 2 above), rather than of the minimal one. Both in Kusuoka and Paclet the closedness of the forms is a consequence of a strict positivity condition of ν.

2) Correspondingly as in the finite dimensional case the question arises of the relation between the different Dirichlet forms. This is as of yet less clarified than in finite dimensions. A strong uniqueness result (essential self-adjointness of the self-adjoint operator associated on E, restricted on FC^∞) is proven in [7], [44] for the case where ν is Gaussian (this includes the case where ν is the "time zero measure" for the free quantum fields).
Recently Takeda [49] has proven a result concerning "Markovian uniqueness", for the case where $\mathcal{H} = L^2(\mathbb{R})$, ν is absolutely continuous with respect to Wiener measure (on $B = C_o[0,1]$).
The problem of uniqueness is a very important one concerning the construction of canonical models of quantum fields, see below.

3) In all cases discussed above where Dirichlet forms have been constructed one gets from them Markov symmetric semigroups P_t in $L^2(\nu)$ (this is actually the main purpose of constructing Dirichlet forms). From these one can always

get symmetric Markov processes with invariant measure ν. The construction of nice realizations of the processes ("diffusions", weak solutions of the stochastic differential equation $d X_t = \beta(X_t)dt + dw_t$, with dw_t the standard Brownian motion associated with $Q \subset \mathcal{H} \subset Q'$ and $\beta(q)(\xi)$ for $q \in Q$, $\xi \in Q'$ s.t. $\beta(q)(\xi) = -(q \cdot \nabla)^* 1(\xi))$ is more complex. It has been undertaken in [45], in the framework of Albeverio, Høegh-Krohn, and in [37], in the framework of Kusuoka (under the assumption that the injection $\mathcal{H} \hookrightarrow B$ is compact and B is either a Hilbert space or a Banach space with some additional assumptions). In particular Kusuoka constructs a diffusion process on the space B.

Potential theory can be discussed in all frameworks, in particular [47] contains a detailed discussion of capacibility of analytic sets.

4) Ergodicity questions have been discussed in the framework of Albeverio - Høegh-Krohn in [7], [44]. These questions should be relevant in connection with Parisi-Wu's approach to stochastic quantization for quantum fields, but the detailed applications have not yet been worked out. The problem of nodes for the finite dimensional densities of ν and its relations with questions of barriers for the corresponding process and quantum field theoretical tunneling could probably be discussed in the framework of [7], [44], [45].

5) Since in infinite dimensions there is no natural Lebesgue reference measure, the semigroup T_t is a purely formal object (but yet suggestive, e.g. in problems of quantum fields, actually non standard analysis gives a meaning to it, see [50]). In applications to quantum fields, one takes $L^2(\nu)$ as the physical Hilbert space and the generator H of $P_t = e^{-tH}$ as the Hamiltonian.
Smoothness property of the energy eigenfunction are statements about eigenfunctions of H, in particular the finite dimensional elliptic result that the lowest end of the spectrum consists of an isolated eigenvalue with strictly positive eigenfunction corresponds here to the statement that the measure ν has a strictly positive smooth density with respect to finite dimensional subspaces, in a sense made precise in [44]. It has been proven to hold in models of

quantum fields in [44].

6) The approximation problem is a hard one already in finite dimensions, and has not been studied in details in the infinite dimensional case (it should here be thought of as the problem of approximating the process associated with P_t by "finite dimensional" processes); some information is of course already available from the construction of the Dirichlet forms and the corresponding semigroup and can be extracted from [7], [44], [37], [47,48].

An interesting class of examples is furnished by quantum field theory.

A) Let us first consider the case of the so called "free quantum fields". In this case $\mathcal{H} = L^2(\mathbb{R}^s)$, with s the dimensionality of space (for the physical space we have s=3). Q can be taken to be $\mathcal{S}(\mathbb{R}^s)$ (real Schwartz functions), $Q' = \mathcal{S}'(\mathbb{R}^s)$ (tempered distributions).
ν is the standard Gauss measure with mean zero and covariance $(-\Delta+m^2)^{-1/2}$, m some positive constant. In this case the associated stochastic differential equation is
$dX_t = \beta(X_t)dt + dw_t$, with w_t Brownian motion associated with $(Q,\mathcal{H},Q')$ (just the process with mean zero and increments such $E(e^{i<q,(w_t-w_s)>}) = \exp(-\frac{1}{2}\|q\|^2)$, $\forall\ q \in Q$, with $\|\ \|$ the norm in $L^2(\mathbb{R}^s)$, $<q,q'>$ the pairing of $q \in Q$, $q' \in Q'$) and $\beta(q)(\xi) = -<\sqrt{-\Delta+m^2}\, q,\xi>$, for all $q \in Q$, $\xi \in Q'$).
Let E the Dirichlet form associated with ν as discussed in 1) and $P_t = e^{-tH}$, $t > 0$ the corresponding semigroup. Then $H \upharpoonright FC^\infty(Q') = (-\frac{1}{2}\Delta - \beta\cdot\nabla) \upharpoonright FC^\infty(Q')$, with $\Delta \equiv \Sigma(e_i\cdot\nabla)^2$, $\beta\cdot\nabla \equiv \Sigma\beta(e_i)(\cdot)(e_i\cdot\nabla)$, is essentially self-adjoint in $L^2(\nu)$, as proven in [7], [44] (the proof only uses that ν is Gaussian and $\beta \in L^2(\nu)$ i.e. $(q\cdot\nabla)^*\ 1 \in L^2(\nu)$ for any basis vector $q \in Q$ in $\mathcal{H}$; it works for any Gaussian measures with covariance A^{-1} s.t. A is trace class).
Thus in particular all Dirichlet forms extending E_o coincide.

There is then only one contraction semigroup in $L^2(\nu)$ with generator coinciding on FC^∞ with $H \restriction FC^\infty$. The Ornstein-Uhlenbeck process X_t has nice realizations as a diffusion process, by the results of [7] and [44]. It is ergodic, zero being a simple isolated eigenvalue of H. ν has strictly positive and smooth density with respect to finite dimensional subspaces.

B) The second class of examples we shall consider is the one given by interacting quantum fields in one space dimension (s=1).
Let $\mathcal{K}$, Q, Q' as in A) and let $\nu = \nu_\ell$, $0 \leq \ell \leq \infty$, where ν_ℓ is the "ground state measure" of model of quantum fields in one space dimension, with interaction in $(-\ell,+\ell)$. For $\ell < \infty$ one speaks of "space cut-off interaction", ν_ℓ is defined, as described in [44], by $\nu_\ell = \Omega_\ell^2 d\nu_0$, with ν_0 the measure as in A) and $\Omega_\ell \in L^1(Q',\nu_0)$ s.t. $\exists$ $E_\ell \in \mathbb{R}$ with $(H_0+V_\ell)\Omega_\ell = E_\ell\Omega_\ell$, $E_0 = \inf \sigma(H_0+V_\ell)$, where $\sigma(\cdot)$ means spectrum and V_ℓ certain function on Q', $V_\ell \in L^p(Q',\nu_0)$ for all $1 \leq p < \infty$ (additive in ℓ). V_ℓ is the space cut-off interaction. For $\ell = \infty$, ν_ℓ is defined as the restriction to the σ-algebra "associated with time zero fields" of the "Euclidean measure" describing a Euclidean quantum field theory in 2 space time-dimensions, see II. In a large class of models [44] one can show ν_ℓ, $1 \leq \ell \leq \infty$ yields a Dirichlet form E, hence by the construction of 3), a symmetric Markov semigroup and a corresponding diffusion process. ν_ℓ is smooth strictly positive ergodic, as shown in [44], which also contains further information on this situation. We shall add a few words on these models in Sect. II.

C) The models considered by Parisi and Wu and others, in the so called method of "stochastic quantization", at least with regularizations, can be treated as processes associated with Dirichlet forms associated with $\mathcal{S}(\mathbb{R}^{s+1}) \subset L^2(\mathbb{R}^{s+1}) \subset \mathcal{S}'(\mathbb{R}^{s+1})$ of the general type discussed here, however with a different kind of ν than the fixed time Euclidean measures discussed in above examples A), B). It would certainly be worthwhile to carry through in details these constructions in the above framework.

II. The relation of the Dirichlet forms and processes of Sect. I with homogeneous Markov random fields.

The theory in Sect. I was, but for the discussion of examples connected with quantum fields, general, in particular Q, $\mathcal{H}$, Q' did not have to consist of functions or distributions in $\mathcal{S}'(\mathbb{R}^s)$. In this section however we shall look upon the processes (X_t,ν) of Sect. I as "marginal processes" obtained from a random field (X,μ) on $\mathbb{R}^d$, $d = s+1$. Interesting examples are furnished by the generalized random fields of Euclidean quantum field theory. However the construction is more general.
Let μ be a measure on $\mathcal{S}'(\mathbb{R}^d)$ (or more generally some space of functions or distributions over $\mathbb{R}^d$), with a support such that for $\xi \in \text{supp}\,\mu$ one can define $<\xi,\rho>$ as an element in $L^2(\nu)$, for all Borel measures ρ on $\mathbb{R}^d$ s.t. $\int d\rho(x)d\rho(y)(-\Delta+m^2)^{-1} < \infty$ for some $0 \leq m$ ($m > 0$ for $d=1,2$). (Such measures exist: e.g. μ the Gaussian measure with mean zero and covariance $(-\Delta+m^2)^{-1}$). Let for any closed measurable $\Lambda \subset \mathbb{R}^d$, $\sigma(\Lambda) = \sigma(<\xi,\rho> \in L^2(\nu)$, $\text{supp}\,\nu \subset \Lambda)$. In particular we have for $\rho(x) = \delta_t(x^1)\varphi(x^2,\ldots,x^d)$ that $<\xi,\rho> \in L^2(\nu)$ and $\sigma(C_t)$ with $C_t = \{x \in \mathbb{R}^d,\ x^1 = t\}$ is well defined. Let $\nu_C = \mu \restriction \sigma(C)$. We assume there exists a structure $(Q,\mathcal{H},Q')$ as in Sect. I s.t. ν_C is a probability measure on Q' and such that $E_0(f,f) = \frac{1}{2}\int(\nabla f,\nabla f)_H d\nu_C$ in $L^2(\nu_C)$, $f \in FC^\infty$, is closable. Then as in Sect. I we get by closure the minimal Dirichlet form $E(f,f)$ and the corresponding semigroup $P_t = e^{-tH}$. Suppose μ is invariant under the transformations U_t of $\mathcal{S}'(\mathbb{R}^d)$: $(U_t\xi)\ (x_1,\ldots,x^d) = \xi(x^1-s,\ x^2,\ldots,x^d))$ induced by the translations τ_t: $x \to (x^1+s,\ x^2,\ldots,x^d)$ in $\mathbb{R}^d$ along the axis orthogonal to C_t, $s \in \mathbb{R}$. Suppose also μ has the global Markov property with respect to C_t, for all $t \in \mathbb{R}$, in the sense that $E(f_+f_-|\sigma(C_t)) = E(f_+|\sigma(C_t))E(f_-|\sigma(C_t))$ $E(\ ,\ )|\sigma(C_t))$ meaning conditional expectation, for all $f_\pm$ bounded and $\sigma(\{x^1 > t\})$ resp. $\sigma\{\{x^1 < t\})$-

measurable. In this case one says that the generalized field (μ,X), with $X_\varphi(\xi) = \langle\xi,\varphi\rangle$, $\varphi \in \mathcal{S}(\mathbb{R}^d)$, has the global Markov property with respect to half-planes [51].

We have that $E(.|C_o)U_tE(.|C_o) \upharpoonright L^2(\nu_{C_o})$ is a Markov semi-group P_t on $L^2(\nu_{C_o})$, with non negative infinitesimal generator H.

The semigroup is symmetric iff μ is invariant under reflections, in the sense of being invariant with respect to the transformation R induced on $\mathcal{S}'(\mathbb{R}^d)$ by $\rho: x \to (-x^1,x^2,\ldots,\ )$ on $\mathbb{R}^d$.

In the latter case H is self-adjoint and

$$(H^{1/2}f,H^{1/2}f)_{L^2(\nu_{C_o})} = \frac{1}{2}\int(\nabla f,\nabla f)_H d\nu_{C_o}$$

for all $f \in FC^\infty$.

H is thus a self-adjoint extension in $L^2(\nu_{C_o})$ of

$-\frac{1}{2}\Delta - \beta - \nabla \upharpoonright FC^\infty$, with $\beta(q)(q') = -(q,\nabla)^* 1(q')$, $q \in Q$, $q' \in Q'$.

In particular ν_{C_o} gives rise to a closable E_o, in the notation of Sect. I, and the Dirichlet form $(H^{1/2},H^{1/2})_{L^2(\nu_{C_o})}$ is a closed extension of E_o.

This then gives an interesting connection between random fields on $\mathcal{S}'(\mathbb{R}^d)$ (translation and reflection invariant) and symmetric Markov processes on Q'. In [44] we discussed a large class of examples where μ_ℓ is the Euclidean measure of quantum fields over $\mathbb{R}^2$ (d=2) with space cut-off so that the Euclidean interaction is in $\mathbb{R} \times (-\ell,+\ell) \subset \mathbb{R}^2$, $0 \leqq \ell \leqq \infty$, in which all hypothesis are satisfied, with $\mathcal{H} = L^2(\mathbb{R})$, $Q = \mathcal{S}(\mathbb{R})$, $Q' = \mathcal{S}'(\mathbb{R})$.

For $\ell < \infty$ all models which have been discussed in the literature have the required global Markov property. For $\ell = \infty$ by now 2 large classes of globally Markov Euclidean fields are known, those with trigonometric interactions [52] and those with exponential interactions [53], [54].

These are then Markov fields with marginals (in the above sense) which are symmetric Markov processes associated with Dirichlet forms of the local type, reducing to the above basic energy form $E_0(f,f)$ on smooth cylinder functions.

There is however a big open problem in this area, namely to prove uniqueness results for the associated semigroups, of the type of problem 2) discussed above in Sect. I and II. The only results up to now, in the infinite dimensional case, are those of Takeda [49], who however treats a mathematical example with ν_{C_0} replaced by a measure absolutely continuous with respect to Wiener measure for Brownian motion.

In the proof of the global Markov property of the Euclidean fields [52] - [54] a solution of the Dirichlet problem with distributional data is involved. Recently a very general treatment of this problem has been given by Röckner, using among other techniques modern tools of axiomatic potential theory. The discussion on Gibbs states, initiated in [52], has also been extended considerably, to a theory of specifications in the sense of Preston.

Let us also remark that the situation discussed above with μ the Euclidean global Markov measure of quantum fields over $\mathbb{R}^2$ has other fascinating features, due to its full invariance with respect to all transformations induced on $\mathcal{S}'(\mathbb{R}^d)$ from Euclidean transformations in $\mathbb{R}^d$. In fact it gives rise to a canonical relativistic theory, as discussed in [21], [23], [46]. This means in particular that there is a representation of the Lorentz group in $L^2(\nu_{C_0})$, with infinitesimal generator of Lorentz transformation also given by a Dirichlet form, at least on FC^∞. There is an operator, canonically conjugate to the time zero field operator $X_\varphi(q') = \langle\varphi,q'\rangle$, $\varphi \in Q = \mathcal{S}(\mathbb{R}^s)$, $q' \in \mathcal{S}'(\mathbb{R}^s)$, in the sense of quantum mechanics (Weyl canonical commutation relation) and one has the "correct equation of motion".

We close this section by mentioning that there is another area in which Dirichlet forms on finite dimensional spaces have already played a role and by the relation through associated random fields the infinite dimensional Dirichlet forms can come to play a role. This is the area of unitary irreducible representations of "gauge groups" i.e. groups of mappings from a manifold into a Lie group. For work in this direction see [21], [23], [62]-[64].

III Markov cosurfaces - a more dimensional analogue of Brownian motion with values in a group

In Sect. I we studied symmetric Markov processes with values in infinite dimensional spaces and in Sect. II we established their connection with Markov random fields. In as much as Markov random fields are extensions from the case of one dimensional time to the case of multidimensional time the consideration of stochastic Markovian mappings from $\mathbb{R}^d$ (or a manifold) into a group G can be looked upon as a non commutative extension of Markov random fields and the associated Markov processes with infinite dimensional state space. One such extension has been studied in details in [21], [23], [62] - [64] It uses essentially methods of the theory of representations of topological groups. In this lecture we shall not discuss this extension, which has already been reviewed elsewhere [23], [62]. We shall rather concentrate on a new extension which is actually framed in terms of random Markovian maps from hypersurfaces to Lie groups ("cosurfaces").

For simplicity we shall first illustrate the concepts in the discrete case in which $\mathbb{R}^d$ is replaced by a lattice $\delta Z^d \equiv M_\delta$, for some $\delta > 0$. We shall call <u>cell</u> γ of M_δ any hypercube $\{x \in \mathbb{R}^d | n_i \delta \leq x_i \leq (n_i+1)\delta,\ i=1,2,\dots,d\}$, for some $n_i \in Z$.

We call <u>face</u> of such a cell γ of M_δ any hypersurface $\{x \in M_\delta | x_i = n_i \delta\}$ or $\{x \in M_\delta | x_i = (n_i+1)\delta\}$ for some i.

Clearly γ has 2d faces, exactly two orthogonal to any given basic vector of $\mathbb{R}^d$. We shall consider oriented cells, the orientation of a given cell being fixed by the one of the <u>basic</u> cell $\{x \in \mathbb{R}^d | 0 \leq x_i \leq \delta,\ i=1,\dots,d\}$, which is the one given by the basic vectors $e_1,\dots,e_d$ in $\mathbb{R}^d$. A face F belonging to a given cell γ, so that $F \subset \partial\gamma$, has then a corresponding orientation.

We shall call H_δ the family of all finite unions of oriented faces of cells of M_δ.

For $S_1, S_2 \in H_\delta$ we define the product $S_1 \cdot S_2$ for d = 2 if the end point of S_1 equals the start point of S_2 resp., for $d > 2$,

if $S_1 \cap S_2$ is (d-2)-dimensional with opposite orientation.
For such S_i, i=1,2 the product $S_1 \cdot S_2$ is by definition $S_1 \cup S_2$ as a set and its orientation is the one inherited from those of S_1 and S_2.
For $S_1,\ldots,S_n \in H_\delta$ we give a recursive definition of $S_1,\ldots,S_n$ as $(S_1,\ldots,S_{n-1}) \cdot S_n$, whenever this is defined. We denote by Σ_δ the set of all products of the form $S_1,\ldots,S_n$ with $S_i \in H_\delta$.
For $S \in H_\delta$ we define S^{-1} as the element of H_δ equal to S as a set, but with opposite orientation.

Rem.: For d = 1, H_δ is simply identifiable with the set of all points of M_δ. In the theory of lattice gauge fields, d=2 the cells are called "plaquettes" and the faces are called "links" or "bonds".

A cosurface C on M_δ with values in a group G is a map C from Σ_δ into G with

1) $C(S^{-1}) = (C(S))^{-1}$
2) $C(S_1 \cdot S_2) = C(S_1)C(S_2)$.

For d = 2 a cosurface is also called a "multiplicative curve integrals". We use here the non commutative notation of product. If G is abelian of course $C(S)^{-1}$ should be understood as $-C(S)$ and $C(S_1) \cdot C(S_2)$ as $C(S_1) + C(S_2)$.
Let $\Gamma_{\delta,G}$ be the set of all cosurfaces on M_δ. $\Gamma_{\delta,G}$ has a natural measurable structure, whenever G has a measurable structure, namely the one making all $S \to C(S)$, $S \in \Sigma_\delta$ measurable.

A stochastic cosurface C on M_δ is a measurable map from some probability space $(\Omega, \mathcal{A}, P)$ into $\Gamma_{\delta,G}$. For d = 1 a stochastic cosurface is just a G-valued stochastic process indexed by δZ, thus a stochastic cosurface is an extension of the concept of stochastic process when the "time points" $t \in \delta Z$ are replaced by d-1-dimensional hypersurfaces S of $M_\delta = \delta Z^d$.

We shall call complex any ordered n-tuple $K=\{S_1,\ldots,S_n\}$ $S_i \in \Sigma_\delta$, for some n. For d=1, K is just an n-tuple of

points. We shall use the notation $C(K)=\{C(S_1),\ldots,C(S_n)\}$ and look upon $C(K)$ as an element in G^K.
Two stochastic cosurfaces C, $\tilde{C}$ are called equivalent if the image probability measures under C, $\tilde{C}$ are equivalent as measures on G^K. We shall henceforth identify equivalent stochastic cosurfaces.

In the following we shall see that it is possible to define a probability measure on a suitable class of complexes. In fact the probability measure is first defined on regular complexes $K = \{S_1,\ldots,S_n\}$ which are <u>regular and saturated</u> in the sense that $S_i \cap S_j \subset \partial S_i \cap \partial S_j$ whenever $i \neq j$, $S_i, S_j \in K$ (regularity) and there exists a partition $D_K = \{A_1,\ldots,A_m\}$, for some $A_i \subset \mathbb{R}^d$ into connected and simply connected subsets A_i of $\mathbb{R}^d$ s.t. $\mathbb{R}^d$ is the union of the A_i and if $A_i \cap A_i \neq \emptyset$ for some $i \neq j$ then either $A_i \cap A_j$ is d-2-dimensional or else $A_i \cap A_j = \bigcup_k S_k^{i,j}$, for some $S_k^{i,j} \in \{S_1,\ldots,S_n\}$ and $\bigcup_{i\neq j} A_i \cap A_j = \bigcup_{i=1}^n S_i$. Assume for simplicity G is compact with countable base. The probability measure will be defined as a projective limit, starting from a definition on regular saturated complexes, which is given in terms of the Haar measure dx on G and a Markov convolution semigroup Q_t, $t \geq 0$ on G (semigroup of Markov kernels on G), with density $Q_t(x)$ s.t. $Q_t(x) \geq 0$, $\int Q_t(xy^{-1})dy = 1$, $\int Q_s(xy^{-1})Q_t(yz^{-1})dy = Q_{s+t}(xz^{-1})$ and $Q_t(x) \to \delta_e(x)$ weakly as $t \downarrow 0$, with δ_e the Dirac measure at the origin $e \in G$. If G is non abelian (which is allowed for $d = 2$) we also assume Q_t invariant in the sense $Q_t(xy) = Q_t(yx)\ \forall x,y \in G$.
For any regular saturated complex $\tilde{K} = \{S_1,\ldots,S_n\}$ we define a probability measure $\mu_{\tilde{K}}^Q$ on $G^{\tilde{K}}$ by

$$d\mu_{\tilde{K}}^Q(C(\tilde{K})) \equiv Z^{-1} \prod_{A \in D_{\tilde{K}}} Q_{|A|} \Big(\prod_{S \in \partial A \cap \tilde{K}} C(S)\Big)dC(S_1)\ldots dC(S_n),$$

where ∂A is the topological boundary of A in $\mathbb{R}^d$, oriented with the normal of A pointing out of A. $|A|$ is the Lebesgue volume of A and we set $Q_\infty = 1$. The product is an ordered product along the oriented boundary ∂A of A. For any complex $K = \{S_1,\dots,S_n\}$ contained in a regular saturated complex $\tilde{K} = \{\tilde{S}_1,\dots,\tilde{S}_n\}$ (in the sense that all the elements in K can be written as a product of the element of K or their inverses) we define $d\mu_K^Q\ (C(K) \equiv \int d\mu_{\tilde{K}}^Q\ (C(\tilde{K}))\ \Pi\ dC(\tilde{S}_i)$, the integral being on the $\tilde{S}_i$ not appearing in K and such that for $S = \tilde{S}_1,\dots,\ \tilde{S}_k$ we have $C(S) = C(\tilde{S}_1),\dots,C(\tilde{S}_k)$.

μ_K^Q is then a measure on the set of all cosurfaces on K. The semigroup property of Q together with the properties of cosurfaces yield that (μ_K^Q, K) form a projective system on G^K, if G is abelian in case $d > 2$. At least when G has a countable base there is no problem then to get a probability space $(\Omega, \mathcal{A}, P)$ and a stochastic cosurface C on it s.t. the image of P under C(K) is μ_K^Q. $(\Omega, \mathcal{A}, P, C)$ is unique up to equivalence. Moreover it is possible to show that C is a <u>Markov cosurface</u> in the following natural sense.

For Λ a subset of M_δ, we shall denote by $\Sigma(\Lambda)$ the σ-algebra generated by all stochastic variables $C(S)$, $S \in \Sigma_\delta$, $S \subset \Lambda$. We say that the stochastic cosurface C has the Markov property i.e. is a Markov cosurface if for any regular complex $K = \{S_1,\dots,S_m\}$ with the properties that $S_{i_1} \cup \dots \cup S_{i}$, $S_{i_k} \in K$, $n=1,\dots,\ell$ splits $\mathbb{R}^d - K$ into 2 connected components $M_K^\pm$ s.t. some of the S_i are in M_K^+ and some are in M_K^- and $E(f_+ f_- | \Sigma(S)) = E(f^+|\Sigma(S))E(f^-|\Sigma(S))$ for any $f^\pm$ which are $\Sigma(M_K^\pm \cup S)$-measurable and bounded. E.g. for $d=2$, $m=3$, $\ell=1$, $i_1 = 2$, $S_2 = \{(x^1,x^2) \in \mathbb{R}^2 | x^2 = 0\}$ this is conditional independence of the curve variables $C(S_i)$, $i = 1$ resp. 3 in $x^2 > 0$ resp. $x^2 < 0$, given C(S). That $(\Omega, \mathcal{A}, P, C)$ has this (global) Markov property is a simple verification from the definition. Hence we have the following

<u>Theorem 1</u> Let G be a compact group with countable base, abelian for $d > 2$. Then μ_K^Q as defined above is a projective

family of probability measures. There exists a probability space $(\Omega,\mathcal{A},P)$ and a stochastic cosurface C s.t. μ_K^Q are the finite dimensional distributions (marginals) of P. $(\Omega,\mathcal{A},P,C)$ is unique up to equivalence. C has the global Markov property.

If we wish to stress the dependence of the Markov cosurface $(\Omega,\mathcal{A},P,C)$ on Q, the volume measure $|A|$ on $\mathbb{R}^d$ and the orientation σ on $\mathbb{R}^d$ we write $C^{Q,\lambda,\sigma}$ for C. Using the invariance of $|A|$ under isometries of $\mathbb{R}^d$ (i.e. piecewise smooth transformations of $\mathbb{R}^d$ leaving λ and σ invariant) and the definition of μ_K^Q we have easily that $C^{Q,\lambda,\sigma}$ is invariant under isometries of $\mathbb{R}^d$ which carry M_δ into itself. Moreover $C^{Q,\lambda,\sigma} = C^{\tilde{Q},\lambda,-\sigma}$ with $\tilde{Q}_t(x) \equiv Q_t(x^{-1})$; in particular if $\tilde{Q}_t = Q_t$, i.e. Q_t is reflection symmetric, then C is independent of σ.

Remark. Under a certain "joint measurability condition" (in ω and S) for a stochastic Markov cosurface $C(S,\omega)$ invariant under isometries of $\mathbb{R}^d$ one can prove a converse of the above result, see [67].

It is not difficult to extend all the preceding results to the case where M_δ is replaced by an oriented Riemannian manifold of dimension d. H_δ is then the family H_M of all oriented piecewise smooth connected (d-1)-dimensional hypersurfaces in M which are closed sets and do not have self-intersections. Σ_M is then defined accordingly as Σ_δ, with H_δ replaced by H_M. The set $\Gamma_{M,G}$ of all cosurfaces on M replaces then $\Gamma_{\delta,G}$. All other definitions hold with M_δ replaced by M (in the definition of regular saturated complex we also replace, of course, $\mathbb{R}^d$ by M). In the definition of $\mu_{\tilde{K}}^Q$ and μ_K^Q we replace the Lebesgue volume $|A|$ by $\lambda(A)$, where λ is the Riemannian volume on M. In the discussion about the dependence of $C^{Q,\lambda,\sigma}$ on λ,σ the isometries should be now understood as piecewise smooth transformations of M leaving λ and σ invariant.

The Theorem and the Remark following it hold then for this situation, yielding a non commutative (G-valued) analogue of a multidimensional time homogeneous Markov field (which can be looked upon, as described in Sect. II, as a symmetric Markov process, given by a Dirichlet form, on an infinite dimensional space). We shall call the object of the Theorem in this case a Markov cosurface over M, with values in G.

We shall remark that it is possible to look upon the Markov cosurface $(\Omega, \mathcal{A}, P, C)$ over $M = \mathbb{R}^d$ as a limit of Markov cosurfaces over a lattice $M_\delta = \delta Z^d$ as $\delta \downarrow 0$.
In fact, let Q_t be a convolution semigroup of Markov kernels on G (compact, and abelian in case $d > 2$).
Let $K = \{S_1, \ldots, S_m\}$, $S_i \in \Sigma_\delta$ be any complex over M. Define μ_K^Q using $Q = Q_t$, starting for (1), as above, with $Q_{|A|}$ replaced by $Q_{\beta|A|}$, with $|A|$, as before, the Riemann-Lebesgue volume of A.
Using the Kolmogorov construction leading to Theor. 1 we get a probability measure P_δ as projective limit of the μ_K^Q (we write δ to underline the dependence on δ). Let $\delta = \delta(n) = \delta_o/2^n$, for some fixed $\delta_o > 0$ and with $n \in \mathbb{N}$. We can consider $P_{\delta(n)}$ as a measure on $(\Omega_M, \mathcal{A}_M)$, the underlying measurable space to the measure P_Q of the remark following Theor. 1, by using the natural embeddings of M_δ and Σ_δ into $M = \mathbb{R}^d$ resp. Σ_M, and we then get that $P_{\delta(n)}$ converges weakly as $n \to \infty$ if β is chosen to be n dependent, namely $\beta = 2^{nd}$ (the limit probability measure is then the one $(\Omega_Q, \mathcal{A}_Q, P_Q)$ obtained from Theor. 1 with $Q = Q_t$).

For $d = 2$ there is an interesting connection between the measures P_δ resp. P_Q and gauge fields on the lattice M_δ resp. on $\mathbb{R}^2$. To describe this connection, let us introduce, for general d, the probability measure on Markov cosurfaces $C(S)$ attached to complexes $K = \{S_1, \ldots, S_n\}$, S_i cells of M_δ, by

$$d\mu_{\beta,U}^{\delta}(C(K)) = Z^{-1} \exp\left[-\beta \sum_{i=1}^{n} U(C(S_i))\right] dC(S_1) \ldots dC(S_n),$$

where Z is the normalizing factor and $\beta > 0$, U being an invariant $(U(xy) = U(yx), \forall x,y \in G)$ real-valued function on G.
There is a projective limit measure μ_δ such that the marginals of μ_δ are given by $\mu^Q_{\beta,U}$.
For d = 2, G a Lie group, μ_δ is the Gibbs measure describing lattice gauge fields on the lattice M_δ.

In [67], [68] it is shown, that at least when G = U(1) or Z_2, (for arbitrary d), or G = SU(2), for d = 2 and U is chosen as the real part of a character on G, there is a suitable choice $\beta(n)$ of β, depending on n, such that

$$\int f(C(F_1^\delta),\ldots, C(F_K^\delta))\, d\mu^{\delta/2^n}_{\beta(n),U}$$

converges as $n\to\infty$ for any real-valued bounded measurable function f on G^{2dk}, with F_i^δ the faces of a finite set of cells $\gamma_1,\ldots,\gamma_k$ of M_δ, $\delta > 0$ fixed. The limit is then $\int f dP_Q$, with P_Q as in the Remark following Theor. 1, with Q_t being a Markov semigroup of density kernels on G, constructed in a suitable way from U and β. E.g. for d = 2, G=SU(2), U(g) = sint/sint/2, $t \in [0,2\pi)$, $e^{\pm it/2}$ the eigenvalues of $g \in SU(2)$, $\beta(n) = 2^{2n}$ we have

$$Q_t = e^{\frac{t}{2}\Delta}$$, with Δ the Laplace-Beltrami operator on G.

Since for d = 2 the measure μ_δ is the Gibbs state of gauge fields and the above limit $n\to\infty$ describes the operation of "going to the continuum", we see that the measure P_Q describes the "continuum limit of gauge fields". Thus we also see that our Markov cosurfaces can be obtained as limits of cosurfaces attached to lattice gauge fields models, for d = 2, and for arbitrary d > 2 as limit of cosurfaces attached to certain lattice fields μ_δ.

<u>Remark:</u> In the case $M = \mathbb{R}^d$ one can associate to the covariant Markov cosurfaces $(\Omega_Q, \mathcal{A}_Q, P_Q; C)$ a Markov semigroup e^{-tH}, $t > 0$ acting on a Hilbert space of functions. H can be interpreted as the Hamiltonian of a relativistic theory, in which the relativistic quantum fields are associated with

d-1-dimensional hypersurfaces, rather than points. This theory satisfes all postulates ([69]) for a relativistic theory, see [67], [68] for details.

Let us finally consider the case where G is a connected compact Lie group. In this case there is a nice relation of a Markov cosurface $(\Omega_Q, \mathcal{A}_Q, P_Q, C)$ with solutions of stochastic equations of the form $da = \xi$, where a is a d-1-form on M with values in the Lie algebra $\mathfrak{g}$ of G and ξ is an infinitely divisible generalized random field on M with values in $\mathfrak{g}$ This connection is best described in the abelian case $G = T^\nu$ (the ν-dimensional torus), but holds also for G non abelian, see [67].
Let thus $G = T^\nu$ and let C be a stochastic cosurface on a Riemannian manifold M with values on G. Under a general assumption of "joint measurability" it is possible to write $C(S)$, $S \in \sum$ in the form

$C(S) = e^{2\pi i \int_S a}$, with $a = (a^1,\dots,a^\nu)$

a $\mathbb{R}$-valued d-1-form on M ("d-1-vector form"), where $e^{2\pi i \int_S a}$ stands for $(e^{2\pi i \int_S a^1}, \dots, e^{2\pi i \int_S a})$ and the integrals are understood as linear functionals, as currents, in the Rahm's sense). The following theorem then holds

Theorem 2. a) If C is a T^ν-valued, jointly measurable Markov cosurface on M invariant under isometries of M, then $C(S) = e^{2\pi i \int_S a}$, with a solution of the stochastic differential equation $da = \xi dx$, with ξ an infinitely divisible generalized random field on M with values in $\mathbb{R}^\nu$ and dx the Rieman-Lebesgue volume on M.
b) Conversely, given such a ξ, the stochastic equation $da = \xi dx$ has a solution a such that $e^{2\pi i \int_S a}$ is a T^ν-valued invariant Markov cosurface.

The proof of a) uses the fact of the Remark following Theor. 1, by which such a C has distribution $(\Omega_Q, \mathcal{A}_Q, P_Q)$ for some Q. In the case $Q_t = e^{\frac{t}{2}\Delta}$, e.g., we have that $\int_S a^j$ for $S = A$ is distributed as $\int_0^{|A|} X_\tau d\tau$ with X_τ white noise on $\mathbb{R}$.

Thus, for any $\beta \in \mathbb{R}^\nu$, $E(e^{<\int_S a,\beta>}) = e^{-\frac{1}{2}|A|\beta^2}$, with $\beta^2 = <\beta,\beta>$, $<,>$ being the scalar product in $\mathbb{R}^\nu$. But this identifies $< \int_A a,\beta>$ with $<\xi,\beta\chi_A>$, proving $da = \xi$.

<u>Remark</u> There is an extension of this result to G non abelian and to non invariant Markov cosurfaces [67]. Results on continuity properties of $S \to C(S)$ can also be obtained [67], [70].

<u>Remark</u> The connection, for d = 2, with gauge fields can be shortly put in the following form. For $S = \partial\gamma$, γ a cell, C(S) is a quantized Wilson loop for a continuum gauge field theory. Let E be the principal fiber bundle with base space M and fiber G. Let a be the 1-form an M obtained from the Lie-algebra valued 1-form $\tilde{a}$ on E giving a connection on E. Then, for $G = T^\nu$, $C(S) = e^{2\pi i \int_S a}$ is the holonomy operator. Da = da = F is the $G = T^\nu$-Yang-Mills fields strength curvature form, which is the above infinitely divisible generalized random field ξ.

In conclusion, in this section we have sketched an extension to the case of state space a group G of the concept of Markov random field on a manifold M. The extension has also provided the starting point for a stochastic calculus for d - 1 stochastic forms and for an extension of multiplicative stochastic integrals to multiplicative stochastic integrals of currents. Inasmuch as Markov random fields can be looked upon as Markov processes with infinite dimensional state space, our discussion in this section can be also looked upon as a non commutative extension of those made in the previous sections. In all cases we have seen a close interconnection of several problems and we might hope that progress in one domain will have implications also in the other ones.

Acknowledgements

The first author would like to thank Professor Ludwig Arnold and Professor Peter Kotelenez for their kind invitation to the workshop. Useful discussions with A. Kaufmann and M. Koeck in Bochum are gratefully acknowledged, as well as stimulating stays at the Universities of Bielefeld, Bochum, Oslo and at the Centre for Interdisciplinary Research, Bielefeld, the Bielefeld-Bochum Research Center Stochastic Processes (BiBoS) and the Centre de Physique Théorique CNRS, Marseille. The partial financial support of the Volkswagenstiftung and the Norwegian Research Concil for Science and the Humanities is also gratefully acknowledged. We thank Mrs. Mischke and Mrs. Richter for the skilfull typing.

S. Albeverio,
Mathematisches Institut,
Ruhr-Universität, Bochum
and
BiBoS-Research Center of
the Volkswagenstiftung

R. Høegh-Krohn,
Matematisk Institutt,
Universitetet i Oslo
and
Université de Provence
et Centre de Physique
Théorique, Marseille

H. Holden
Matematisk
Institutt
Universitetet
i Oslo

References

[1] M. Fukushima, Dirichlet forms and Markov processes, North Holland Kodansha (1980)

[2] M.L. Silverstein, Symmetric Markov processes, Lect. Notes Maths. 426, Springer, Berlin (1974)

[3] M.L. Silverstein, Boundary theory for symmetric Markov processes, Lect. Notes Maths. 516, Springer, Berlin (1976)

[4] M. Fukushima, Energy forms and diffusion processes, ZiF-Preprint, in preparation (to appear in Proc. Bielefeld, Project 2: Mathematics - Physics, Ed. L. Streit, World Publishing, Singapore)

[5] E. Nelson, Dynamical Theories of Brownian motion, Princeton University Press, 1970; Quantum Fluctuations, Princeton University, in preparation

[6] S. Albeverio, R. Høegh-Krohn, A remark on the connection between stochastic mechanics and the heat equation, J. Math. Phys. 15, 1745-1747 (1974)

[7] S. Albeverio, R. Høegh-Krohn, Quasi invariant measures, symmetric diffusion processes and quantum fields, pp. 11-59 in "Proceedings of the International Colloquium on Mathematical Methods of Quantum Fields Theory", Ed. CNRS (1976)

[8] S. Albeverio, R. Høegh-Krohn, L. Streit, Energy forms, Hamiltonians and distorted paths, J. Math. Phys. 18, 907-917 (1977)

[9] M. Fukushima, A decomposition of additive functionals of finite energy, Nagoya Math. J. 74, 137-168 (1979)

[10] M. Fukushima, On a representation of local martingale additive functionals of symmetric diffusions, pp. 110-118 in Stochastic Integrals, Ed. D. Williams, Lect. Notes Maths. 851, Springer, Berlin (1981)

[11] M. Röckner, N. Wielens, Dirichlet forms - closability and change of speed measure, ZiF-Preprint, to appear in "Infinite dimensional analysis and stochastic processes", Ed. S. Albeverio, Res. Notes Maths., Pitman

[12] K. Rullkötter, U. Spönemann, Dirichletformen und Diffussionsprozesse, Diplomarbeit, Bielefeld; and paper in preparation

[13] J. Brasche, Störungen von Schrödingeroperatoren durch Masse, Diplomarbeit, Bielefeld (1983); Perturbations of Schrödinger Hamiltonians by measures-self-adjointness and lower semiboundedness, ZiF-Preprint (1984) (to appear in J. Math. Phys.)

[14] W. Karwowski, J. Marion, On the closability of some positive definite symmetric differential forms on $C_o^\infty(\Omega)$, CNRS, Marseille, Preprint (1984)

[15] S. Albeverio, R. Høegh-Krohn, L. Streit, Regularization of processes and Hamiltonians, J. Math. Phys. 21, 1636-1642 (1980)

[16] S. Albeverio, S. Kusuoka, L. Streit, Convergence of Dirichlet forms and associated Schrödinger operators, BiBoS-Preprint (1984)

[17] M. Fukushima, On a stochastic calculus related to Dirichlet forms and distorted Brownian motions, Phys. Repts. 77, 255-262 (1981)

[18] I. G. Hooton, Dirichlet forms associated with hypercontractive semigroups, Trans. Am. Math. Soc. 253, 237-256 (1979)

[19] N. Wielens, Eindeutigkeit von Dirichletformen und wesentliche Selbstadjungiertheit von Schrödingeroperatoren mit stark singulären Potentialen, Diplomarbeit, Bielefeld (1983)
On the essential self-adjointness of generalized Schrödingeroperators, J. Funct. Anal., 1984

[20] L. Streit, Energy forms: Schrödinger theory, processes, Phys. Repts. 77, 363-375 (1981)

[21] S. Albeverio, R. Høegh-Krohn, Markov processes and Markov fields in quantum theory, group theory, hydrodynamics and C*-algebras, pp. 497-540 in "Stochastic Integrals", Ed. D. Williams, Lect. Notes Maths. 851, Springer, Berlin (1980)

[22] S. Albeverio, R. Høegh-Krohn, Some remarks on Dirichlet forms and their applications to quantum mechanics and statistical mechanics, pp. 120-132 in Functional Analysis in Markov-Processes, Ed. M. Fukushima, Lect. Notes Maths. 923, Springer, Berlin (1982)

[23] S. Albeverio, R. Høegh-Krohn, Diffusion fields, quantum fields and fields with values in groups in Adv. probability, Stochastic Analysis, Ed. M. Pinsky, Dekker, New York (1984)

[24] R. Carmona, Processus de diffusion gouvernés par la forme de Dirichlet de l'opérateur de Schroedinger, Sém. Prob. XIII, Lect. Notes Maths. 721, Springer, Berlin (1979)

[25] S. Albeverio, F. Gesztesy, W. Karwowski, L. Streit, On the connection between Schrödinger- and Dirichlet forms, BiBoS-Preprint (1984)

[26] M. Nagasawa, Segregation of a population in an environment, J. Math. Biology 9, 213-235 (1980)

[27] S. Albeverio, M. Fukushima, W. Karwowski, L. Streit, Capacity and quantum mechanical tunneling, Comm. Math. Phys. 80, 301-342 (1981)

[28a] S. Albeverio, Ph. Blanchard, F. Gesztesy, L. Streit, Quantum mechanical low energy scattering in terms of diffusion processes, ZiF-Preprint 1984, to appear in Proc. II Enc. Math. Phys., France-Germany, Lect. Notes Maths., Springer (1984)

[28b] S. Albeverio, W. Karwowski, L. Streit, Some remarks on capacity, Green functions and Schrödinger operators, to appear in Ref. 11

[29] M. Fukushima, Markov processes and functional analysis, pp. 187-202 in Proc. Int. Math. Conf., L.H.Y. Chen et al, Edts., North Holland (1982)

[30] M. Fukushima, A note on irreducibility and ergodicity of symmetric Markov processes, pp. 200-208 in "Stochastic processes in quantum hteory and statistical physics", Edt. S. Albeverio et al, Lect. Notes Phys. 173, Springer, Berlin (1982)

[31] E. Carlen, Conservative diffusions and constructive approach to Nelson stochastic mechanics, Princeton Univ. Ph D Thesis (1984)

[32] W.A. Cheng, Sur la méchanique stochastique, RMA Strasbourg (1984)

[33] a) M. Fukushima, On distorted Brownian motions, pp. 255-262 in Ref. 20
b) M. Fukushima, On absolute continuity of multidimensional symmetrizable diffusions, pp. 146-176 in Ref.22

[34] a) H. Nagai, Impulsive control of symmetric Markov processes and quasi-variational inequalities, Osaka J. Math. 20, 863-879 (1983)
b) H. Nagai, Stopping problems of symmetric Markov processes and nonlinear variational inequalities, Preprint 1984

[35] M. Fukushima, A generalized stochastic calculus in homogeneization, pp.42-51 in "Quantum fields - Algebras, Processes", Proc. Symp. Bielefeld, 1978, Ed. L. Streit, Springer, Wien (1980)

[36] H. Föllmer, Dirichlet processes, pp. 476-478 in Ref. 21

[37] S. Kusuoka, Dirichlet forms and diffusion processes on Banach spaces, J. Fac. Sci. Univ. Tokyo 1A29, 79-95 (1982)

[38] M. Fukushima, M. Takeda, A transformation of a symmetric Markov processes and the Donsker-Varadhan theory, Osaka J. Math. 21, 311.326 (1984)

[39] M. Fukushima, M. Okada, On conformal martingales diffusions and pluripolar sets, J. Funct. Anal. 55, 377-388 (1984)

[40] S. Albeverio, Ph. Blanchard, R. Høegh-Krohn, A stochastic model for the orbits of planets and satellites - The Titius-Bode law, Exp. Math. 4, 365-373 (1983)

[41] S. Albeverio, Ph. Blanchard, R. Høegh-Krohn, Diffusions sur une variété Riemannienne: barrières infanchissables et applications, Coll. L. Schwartz, Astérisque 1984

[42] S. Albeverio, Ph. Blanchard, Ph. Combe, R. Høegh-Krohn, R. Rodriguez, M. Sirugue, M. Sirugue-Collin, Magnetic bottles in a dirty environment. A stochastic model for radiation belts, ZiF-Preprint 1984

[43] S. Albeverio, R. Høegh-Krohn, The structure of diffusion processes, Marseille, Preprint

[44] S. Albeverio, R. Høegh-Krohn, Dirichlet forms and diffusion processes on rigged Hilbert spaces, Zeitschr. für Wahrscheinlichkeitstheorie verw. Gebiete 40, 1-57 (1977)

[45] S. Albeverio, R. Høegh-Krohn, Hunt processes and analytic potentials theory on rigged Hilbert spaces, Ann. Inst. H. Poincaré B 13, 269-291 (1977)

[46] a) S. Albeverio, R. Høegh-Krohn, Canonical quantum fields in two space-time dimensions, Preprint
b) S. Albeverio, R. Høegh-Krohn, Topics in infinite dimensional analysis, pp. 278-302 in Lect. Notes Phys. 80, Springer (1978) (Ed. G.F. Dell'Antonio et al)

[47] Ph. Paclet, Espaces de Dirichlet et capacités fonctionnelles sur des triplets de Hilbert-Schmidt, Sém. Paul Krée, 1977-1978 No. 5, 1-36

[48] Ph. Paclet, Espaces de Dirichlet endimension infinie, C.R. Ac.-Sciences Paris Ser. A 288, 981-993 (1979)

[49] M. Takeda, On the uniqueness of Markovian extension of diffusion operators on infinite dimensional spaces, ZiF-Preprint, 1984

[50] S. Albeverio, J.E. Fenstad, R. Høegh-Krohn, T. Lindstrøm Non standard methods in stochastic analysis and mathematical physics, book in preparation (Acad. Press, (1985)

[51] B. Simon, The $P(\varphi)_2$ Euclidean (Quantum) field theory, Princeton Univ. (1974)

[52] S. Albeverio, R. Høegh-Krohn, Uniqueness and the global Markov property for Euclidean fields. The case of trigonometric interactions, Comm. Math. Phys. 68, 95-128 (1979)

[53] R. Gielerak, Verification of the global Markov property in some class of strongly coupled exponential interactions, J. Math. Phys. 24, 347-355 (1983)

[54] B. Zegarlinski, Uniqueness and the global Markov property for Euclidean fields: the case of general exponential interactions, Comm. Math. Phys., to appear

[55] S. Albeverio, R. Høegh-Krohn, Uniqueness and global Markov property for Euclidean fields and Lattice systems, pp. 303-329 in Quantum fields, Algebra Processes, Ed. L. Streit, Springer Verlag, Wien (1980)

[56] a) S. Albeverio, R. Høegh-Krohn, Stochastic methods in quantum field theory and hydrodynamics, Phys. Repts. 77, 193-214 (1981)
b) S. Albeverio, R. Høegh-Krohn, Local and global Markoff fields, Repts. Math. Phys. 19, 225-248 (1984)

[57] M. Röckner, Generalized Markov fields and axiomatic potential theory, Math. Ann. 264, 153-177 (1983)

[58] M. Röckner, Self-adjoint harmonic spaces and Dirichlet forms, Hiroshima Math. J., to appear

[59] M. Röckner, A Dirichlet problem for distributions and the construction of specifications for Gaussian generalized random fields, Mem. Am. Math. Soc., to appear

[60] M. Röckner, Generalized Markov fields and Dirichlet forms, Acta Applic. Math., to appear

[61] a) E.B. Dynkin, Markov processes and random fields, Bull Am. Math. Soc. 3, 975-999 (1980)
b) T. Kolsrud, in preparation

[62] S. Albeverio, R. Høegh-Krohn, D. Testard, Factoriality of representations of the group of paths of SU(n), J. Funct. Anal. 57, 49-55 (1984)

[63] J. Marion, D. Testard, Energy representations of gauge groups coming from foliations, Luminy, Preprint (1984)

[64] S. Albeverio, R. Høegh-Krohn, Diffusion, quantum fields and groups of mappings, pp. 133-145 in Ref. 22

[65] L. Gross, Logarithmic Sobolev inequalities, Amer. J. Math. 97, 1061-1083 (1975)

[66] B. Lascar, Opérateurs pseudo-différentiels en dimension infinie. Etude de l'hypoellipticité et de la résolubilité dans des classes de fontions hölderiennes et de distribution pour des opérateurs pseudo-différentiels elliptiques, J. Anal. Math. 28, 39-104 (1978)

[67] S. Albeverio, R. Høegh-Krohn, H. Holden, in preparation

[68] S. Albeverio, R. Høegh-Krohn, H. Holden, Markov cosurfaces and gauge fields, Acta Phys. Austr. Suppl. 26, 211-231 (1984)
(Stochastic Methods and Computer Techniques in Quantum Dynmaics, Ed. H. Mitter, L. Pittner)

[69] E. Seiler, Gauge theories as a problem of construction quantum field theory and statistical mechanics, Lect. Notes Phys. 159, Springer, Berlin (1982)

[70] A. Kaufmann, in preparation

[71] J. Brasche, BiBoS-Preprint, in preparation

G. DA PRATO

MAXIMAL REGULARITY FOR STOCHASTIC CONVOLUTIONS AND APPLICATIONS TO STOCHASTIC EVOLUTION EQUATIONS IN HILBERT SPACES

1. INTRODUCTION

We are here concerned with the following problem:

$$(1.1)\quad dy = Ay\,dt + \sum_{i=1}^{N} B_i y\, dw_i(t) \quad , \quad y(0) = x$$

where $A, B_1, B_2, \ldots, B_n$ are linear operators (generally unbounded) in a Hilbert space H and $w_1, w_2, \ldots, w_N$ are real Brownian motions in a probability space (Ω, ε, P). When A generates a strongly continuous semi-group in H and $B_1, B_2, \ldots, B_N$ are bounded, problem (1.1) has been extensively studied. In this case Eq. (1.1) is written as an integral equation

$$(1.2)\quad y(t) = e^{tA} x + \sum_{i=1}^{N} \int_0^t e^{(t-s)A} B_i y(s)\, dw_i(s)$$

which is solved by successive approximations (see [1], [2], [5], [6], [7], [14], [15], [19], [20], [21], [22], [23], [24], [28]).
When B_i are unbounded the most used approach is the variational one (see [29]).
If $N = 1$ and $B = B_1$ generate a group it is possible to reduce Eq. (1.1) to a deterministic equation by setting $v = e^{Bw_t} u$. In this case we obtain the problem:

L. Arnold and P. Kotelenez (eds.), Stochastic Space-Time Models and Limit Theorems, 41–52.

$$(1.3)\quad \begin{cases} v'(t,\omega) = e^{-Bw_t(\omega)}\left(A - \frac{B^2}{2}\right)e^{Bw_t(\omega)}\, v(t,\omega) \\ v(0,\omega) = x \qquad\qquad \forall\omega \in \Omega \end{cases}$$

and this problem can be studied via the abstract evolution equations theory. In this paper we will give some results using the regularizing properties of the Stochastic Convolution (see [8] and [9]).
In §2 we shall recall the main properties of Stochastic Convolution, in §3 we shall give some existence results.

2. STOCHASTIC CONVOLUTION

We assume here that A generates an analytic semi-group e^{tA} of negative type and that $f_i \in M_w^2(0,T;H)$, $i = 1,2,\dots,N$. $M_w^2(0,T;H)$ represents the set of all stochastic processes adapted to $w(t) = w_1(t),\dots,w_N(t))$ such that

$$E\int_0^T |g(s)|^2\, dw_i(s) < +\infty \qquad i = 1,\dots,N \ .$$

We set

$$(2.1)\quad X_t = \sum_{i=1}^{N} \int_0^t e^{(t-s)A} f_i(s)\, dw_i(s) \ .$$

X_t is called a <u>stochastic convolution</u>.

In the deterministic case the convolution $e^{tA} * f (f \in L^2(0,T;H))$ has the following regularizing property:

(2.2) $f \in L^2(0,T;H) \Rightarrow e^{tA} * f \in L^2(0;T,D(A))$

where $(e^{tA} * f)(t) = \int_0^t e^{(t-s)A} f(s)ds$.

Thus it is natural to expect that X_t takes values in some interpolation space between the domain $D(A)$ of A and H.

Let us recall the definition of interpolation spaces $D_A(\theta,2)$, $\theta \in]0,1[$ (see [4], [18], [27]). We say that x belongs to $D_A(\theta,2)$ if the following integral is finite:

(2.3) $|x|^2_{D_A(\theta,2)} = \int_0^\infty \xi^{1-2\theta} |A\ e^{\xi A}\ x|^2\ d\xi$.

If A is self-adjoint negative then $D_A(\theta,2)$ is isomorphic to $D((-A)^\theta)$, moreover

(2.4) $|x|^2_{D_A(\theta,2)} = 2^{2\theta-2}\ \Gamma(2 - 2\theta) |(-A)^\theta x|^2_H$

We shall set in the sequel

(2.5) $|x|_{D((-A)^\theta)} = |(-A)^\theta\ x|$

<u>Remark 2.1</u>

If $\theta \in]0,\frac{1}{2}[$ we have

(2.6) $|e^{tA}|_{\mathcal{L}(D_A(\theta,2))} \leq 1$.

Moreover, if $D(A) = D(A^*)$ (A^* being the adjoint of A) there exists an operator A_1 and $\theta \in]0,\frac{1}{2}[$ such that $D_A = D_{A_1}(\theta,2)$; thus in this case there exists an equivalent scalar product in H with respect to which e^{tA} is a contraction

The following result is proved in [8] .

Proposition 2.2

Assume that $\theta \in]0,\frac{1}{2}[$, $f_i \in M^2_w(0,T;D_A(\theta,2))$, $i=1,2,\dots,N$. Then $X \in M^2_w(0,T;D_A(\theta+\frac{1}{2},2))$ and the following inequality holds

$$(2.7)\quad \|X\|^2_{M^2_w(0,T;D_A(\theta+\frac{1}{2},2))} \le \frac{1}{1-2\theta}\sum_{i=1}^{N}\|f_i\|^2_{M^2_w(0,T;D_A(\theta,2))} .$$

If A is self-adjoint we have the result

Proposition 2.3

Assume that A is self-adjoint negative, $\theta \in [0,\frac{1}{2}]$ and $f_i \in M^2_w(0,T;D((-A)^{\theta}))$. Then $X \in M^2_w(0,T;D((-A)^{\theta+1/2}))$ and the following inequality holds

$$(2.8)\quad \|X\|^2_{M^2_w(0,T;D((-A)^{\theta+1/2}))} \le \frac{1}{2}\sum_{i=1}^{N}\|f_i\|^2_{M^2_w(0,T;D((-A)^{\theta}))}$$

Remark 2.4

Assume that the hypotheses of Proposition 2.1 hold. Then since e^{tA} is a contraction in $D_A(\theta,2)$, $X(\cdot,\omega)$ is continuous in $D_A(\theta,2)$ a.e. in Ω (see [20] and [23]). For the same reason if A is self-adjoint or if $D(A) = D(A^*)$, $x(\cdot,\omega)$ is continuos in H a.e. in Ω.
In the general case we do not know if $X(\cdot,\omega)$ is continuous in H. Moreover, if $f_i \in L^p_w([0,T];H)$ with $p > 2$, this is true ([13]).

3. EXISTENCE RESULTS

We are here concerned with the equation

$$(3.1)\quad y(t) = e^{tA}x + \sum_{i=1}^{N}\int_0^t e^{(t-s)A}B_i y(s)\,dw_i(s)$$

where e^{tA} is a strongly continuous semi-group in H.

Setting $z(t) = e^{-\omega t}y(t)$, $A_1 = A - \omega$, Eq. (3.1) reduces to

$$(3.2)\quad z(t) = e^{tA_1}x + \sum_{i=1}^{N}\int_0^t e^{(t-s)A_1} B_i z(s)\, dw_i(s) \quad ,$$

so it is not a restriction to assume e^{tA} of negative type.

Proposition 3.1

Let $\theta \in]0,\frac{1}{2}[$. Assume that:

$$(3.3)\quad \begin{cases} \text{a) } e^{tA} \text{ is an analytic semi-group of negative type and } x \in D_A(\theta,2). \ B_i \text{ maps } D_A(\theta+\frac{1}{2},2) \text{ into } D_A(\theta,2). \text{ Moreover} \\ \text{b) } \exists\, \eta \in [0,1[\text{ such that} \\ \dfrac{1}{1-2\theta}\displaystyle\sum_{i=1}^{N} |B_i z|^2_{D_A(\theta,2)} \le \eta\, |z|^2_{D_A(\theta+\frac{1}{2},2)} \quad \forall z \in D_A(\theta+\tfrac{1}{2},2) \end{cases}$$

then Eq. (3.1) has a unique solution $y \in M_w^2(0,T;D_A(\theta+\frac{1}{2},2))$. Moreover $y(\cdot,\omega)$ is continuous in $D_A(\theta,2)$ for a.e. $\omega \in \Omega$.

Proof

We can write Eq. (3.1) in the form

$$y = e^{tA}x + \Gamma(y) \quad .$$

Let $Z = M_w^2(0,T;D_A(\theta+\frac{1}{2},2))$, then by Proposition 2.2 and 3.1 Γ maps Z into itself and $|\Gamma|_{\mathcal{L}(Z)} = \eta < 1$. Moreover $e^{tA}x \in Z$, as can be easily checked, so that the conclusion follows from the Contraction principle and Remark 2.3

Remark that the solution y of Eq. (3.1) is not, in general, a strong solution of:

$$(3.4)\quad dy(t) = Ay(t)dt + \sum_{i=1}^{N} B_i y(t)\, dw_i(t) \quad , \quad y(0) = x$$

For this we need a stronger hypothesis on B_i. The proof of the following proposition is similar to that of Proposition 3.1.

Proposition 3.2

Let $\theta \in]0,\frac{1}{2}[$. Assume that

$$(3.5)\quad \begin{cases} \text{a)} & e^{tA} \text{ is an analytic semi-group of negative type, } x \in D_A(\theta + \frac{1}{2},2) \\ \text{b)} & B_i \text{ maps } D_A(\theta + 1,2) \text{ into } D_A(\theta + \frac{1}{2},2) \text{ and } \exists\, \eta \in [0,1] \text{ such that} \\ & \frac{1}{1-2\theta} \sum_{i=1}^{N} |B_i z|^2_{D_A(\theta+\frac{1}{2},2)} \le \eta |z|^2_{D_A(\theta+1,2)} \quad \forall z \; D_A(\theta+1,2) \end{cases}$$

then Eq. (3.4) has a unique strong solution y with $y \in M^2_w(0,T;D_A(\theta + 1,2))$. Moreover $y(\cdot,\omega)$ is continuous in $D_A(\theta + \frac{1}{2},2)$ a.e. in Ω.

Let us consider now the case in which A is self-adjoint. Then, arguing as for the proofs of Proposition (3.1) and (3.2) we get the following:

Proposition 3.3

Let $\theta \geq 0$. Assume that

$$(3.6)\quad \begin{cases} \text{a)} & A \text{ is self-adjoint negative, } x \in D((-A)^{\theta}). \\ \text{b)} & B_i \text{ maps } D((-A)^{\theta+\frac{1}{2}}) \text{ into } D((-A)^{\theta}) \text{ and} \\ & \exists\, \eta \in [0,1[\text{ such that} \\ & \frac{1}{2} \sum_{i=1}^{N} |B_i z|^2_{D((-A)^{\theta})} \le \eta |z|^2_{D((-A)^{\theta+1/2})} \quad \forall z \in D((-A)^{\theta+\frac{1}{2}}) \end{cases}$$

then Eq. (3.1) has a unique solution $y \in M^2_w(0,T;D((-A)^{\theta+\frac{1}{2}}))$. Moreover $y(\cdot,\omega)$ is continuous in $D_A(\theta,2)$ for a.e. $\omega \in \Omega$. Finally if $\theta \geq \frac{1}{2}$ then y is a strong solution of Eq. (3.4).

Remark 3.4

Hypothesis (3.3)-c is equivalent to $\gamma_\theta < 1$, where

$$(3.7)\quad \gamma_\theta = \frac{1}{1-2\theta} \operatorname{Sup}\{ \sum_{i=1}^{N} |B_i x|^2_{D_A(\theta,2)} ; |x|_{D_A(\theta+\frac{1}{2},2)} \le 1\}$$

When A is self-adjoint it is not difficult to compute γ_θ whereas in the general case we can only give estimates for γ_θ (see Example 3). For this reason if $A = A_1 + C$, with A_1 self-adjoint and C continuous from $D_A(\theta + \frac{1}{2},2)$ into $D_A(\theta,2)$ it is more convenient to consider, instead to Eq. (3.1), the following one

$$(3.8) \quad y(t)=e^{tA_1}x + \int_0^t e^{(t-s)A_1}Cy(s)ds + \sum_{i=1}^N \int_0^t e^{(t-s)A_1}B_i y(s)dw_i(s)$$

that we can solve under hypotheses (3.6) (with A replaced by A_1).

We shall give now some examples, for sake of simplicity we shall only consider differential operators in one space variable. Many results can be easily generalized for operators in several variables.

<u>Example 3.5</u> (Dirichlet problem)

Let $H = L^2(0,\pi)$, $Au = u_{xx}$, $D(A) = H^2(0,\pi) \cap H_0^1(0,\pi)$, $Bu = bu_x$, with $b \in \mathbb{R}$. Equation (1.1) becomes

$$(3.9) \quad du = u_{xx}\, dt + bu_x\, dw_t \quad , \quad u(0) = u_0$$

In this case A is self-adjoint, moreover we have (see for instance [26])

$$(3.10) \quad D((-A)^\theta) = \begin{cases} H^{2\theta}(0,\pi) & \text{if } \theta \in]0,\frac{1}{4}[\\ \{u \quad H^{2\theta}(0,\pi); u(\cdot)=u(1)=0\} & \text{if } \theta \in]\frac{1}{4},1[\end{cases}$$

Moreover

$$(3.11) \quad |(-A)^\theta u|^2_{L^2(0,\pi)} = \frac{2}{\pi} \sum_{k=1}^\infty k^{4\theta} \left(\int_0^\pi u(x)\sin kx\, dx\right)^2 = |u|^2_{H^{2\theta}(0,\pi)}$$

Fix now $\theta \in [0,\frac{1}{4}[$ then B maps $D((-A)^{\theta+\frac{1}{2}})$ into $D((-A)^\theta)$; moreover if $u \in D((-A)^{\theta+1/2})$

$$(3.12)\quad |u_x|^2_{D((-A)^\theta)} = \frac{2}{\pi}\sum_{k=1}^{\infty} k^{4\theta+2}\left(\int_0^\pi u(x)\cos kx\,dx\right)^2 =$$

$$= \frac{2}{\pi}\sum_{k=1}^{\infty} k^{4\theta+2}\left(\int_0^\pi u(x)\sin kx\,dx\right)^2 = |u|^2_{D((-A)^{\theta+1/2})}$$

it follows

$$(3.13)\quad \frac{1}{2}|Bu|^2_{D((-A)^\theta)} = \frac{b^2}{2}|u|^2_{D((-A)^{\theta+1/2})}$$

so that $\gamma_\theta = b^2/2$.

Thus if $|b| < \sqrt{2}$ and $u_0 \in H^{2\theta}(0,\pi)$, $\theta \in [0,\frac{1}{4}[$ there exists a unique solution $u \in M^2_w(0,T;\ H^{2\theta+1}(0,\pi) \cap H^1_0(0,\pi))$ to Eq. (3.1).

Remark that the condition $|b| < \sqrt{2}$ is the same that appears in the variational theory ([29]). This condition is quite natural, in fact if we consider the equation

$$(3.14)\quad du = u_{xx}\,dt + b\,u_x\,dw_t\ ,\quad u(0) = u_0$$

in $L^2(\mathbb{R})$ we can write down the following explicit solution

$$(3.15)\quad u(t,x) = e^{t(1-b^2/2)D_x^2}\, e^{w_t D_x}\, u_0(x)$$

Thus if $b < \sqrt{2}$ we have no solution and if $b = \sqrt{2}$ the solution does not belong to $H^{2\theta+1}$ (for general u_0).

<u>Example 3.6</u> (Neumann Problem)

Let $H = L^2(0,\pi)$, $A_1 u = u_{xx}$, $D(A_1) = \{u \in H^2(0,\pi);$ $u'(0) = u'(\pi) = 0\}$ $B = b\,u_x$, $b \in \mathbb{R}$. A_1 is self-adjoint and we have

$$(3.16)\quad D((-A_1)^\theta) = \begin{cases} H^{2\theta}(0,\pi) & \text{if } \theta \in]0,\frac{3}{4}[\\ \{u \in H^{2\theta}(0,\pi); u'(0)=u'(\pi)=0\} & \text{if } \theta \in]\frac{3}{4},1[\end{cases}$$

Fix now $\theta \in]\frac{1}{2},\frac{3}{4}[$; then A_1 maps $D((-A_1)^{\theta+1/2})$ into $D((-A)_1^{\theta})$; moreover proceeding as in Example (3.5) we can show that

(3.17) $$\frac{1}{2}|Bu|^2_{D((-A_1)^{\theta})} = \frac{b^2}{2}|u|^2_{D((-A_1)^{\theta+1/2})}$$

Thus if $|b| < \sqrt{2}$, $\theta \in]\frac{1}{2},\frac{3}{4}[$ and $u_0 \in H^{2\theta}(0,\pi)$ there exists a unique strong solution $u \in M^2_w(0,T;H^{2\theta}(0,\pi))$ to the equation

(3.18) $$du = u_{xx}\,dt + b\,u_x\,dw_t$$

(with boundary Neumann conditions)

Example 3.7 (Non variational)

Let $H = L^2(0,\pi)$, $A_2u = du_{xx}$ with $\alpha \in C([0,\pi])$, $D(A_2) = H^2(0,\pi) \cap H^1_0(0,\pi)$. We assume

(3.19) $$0 < \varepsilon \leq \alpha \leq k$$

Let $Bu = bu_x$, $b \in \mathbb{R}$, $D(B) = H^1_0(0,\pi)$. Remember that

(3.20) $$D_{A_2}(\theta,2) = D_A(\theta,2)$$

where A is the operator defined in Example (3.5).
Let us compute γ_θ; we have

(3.21) $$|u|^2_{D(A_1)} \leq k^2|u|^2_{D(A)} \leq \frac{k^2}{2}|u|^2_{D(A_1)}$$

by interpolation we get

(3.22) $$|u|_{D_{A_1}(\theta,2)} \leq k^\theta|u|_{D_A(\theta,2)} \leq (\frac{k}{\varepsilon})^\theta|u|_{D_{A_1}(\theta,2)}$$

Using (3.12) we have

$$|Bu|^2_{D_{A_1}(\theta,2)} \leq k^{2\theta}|Bu|^2_{D_A(\theta,2)} = k^{2\theta}2^{2\theta-2}\Gamma(2-2\theta)|Bu|^2_{D((-A)^\theta)}$$

$$\leq k^{2\theta}2^{2\theta-2}\Gamma(2-2\theta)b^2|u|^2_{D((-A)^{\theta+1/2})} = \frac{1-2\theta}{2}k^{2\theta}b^2|u|^2_{D_A(\theta+\frac{1}{2},2)}$$

$$\leq \frac{1}{2}k^{2\theta}(1-2\theta)^{-2\theta-1}b^2|u|^2_{D_{A_1}(\theta+\frac{1}{2},2)}$$

thus

$$\gamma_\theta \leq \frac{b^2}{2}\frac{k^{2\theta}}{\varepsilon^{2\theta+1}} \tag{3.23}$$

and if $|b| < 2\frac{\varepsilon^{\theta+1/2}}{k^\theta}$ we can solve Eq. (3.1).

G. Da Prato
Scuola Normale Superiore
Pisa, Italy

REFERENCES

[1] Arnold, L., Curtain, R., Kotelenez, P.: Nonlinear Stochastic Evolution Equations in Hilbert spaces, Report n° 17, Bremen University (1974).

[2] Balakrishnan, A.: Stochastic Bilinear partial Differential Equations, U.S. Italy Conference on Variable Structure Systems, Oregon (1974).

[3] Bensoussans, A., Temam, R.: Equations aux dérivées pattielles stochastiques non linéaires (1) Isr. J. Math. 11 n° 1, 95-129 (1972).

[4] Butzer, P., Berens, H.: Semigroup of Operators and Approximation, Springer-Verlag (1967).

[5] Chojnowska-Michalik, A.: Stochastic Differential Equations in Hilbert Spaces and some of their applications, Thesis, Institute of Mathematics, Polish Academy of Sciences, 1977.

[6] Curtain, R.F.: Stochastic Evolution Equations with general white noise disturbance. J. Math. Anal. Appl. 60 (1977) 570-595.

[7] Curtain, R.F., Pritchard, A.J.: Infinite Dimensional Linear Systems Theory, Springer-Verlag (1978).

[8] Da Prato, G.: Regularity results of a convolution stochastic integral and applications to parabolic stochastic equations in a Hilbert space, Conference del Seminario di Matematica dell'Università di Bari (1982).

[9] Da Prato, G.: Some results on linear stochastic evolution equations in Hilbert spaces by the semi-groups methods, Stochastic Analysis and Applications 1, 1, 57-88 (1983).

[10] Da Prato, G., Grisvard, P.: Equations d'évolution abstraites non linéaires de type parabolique, Annali di Matematica pura e applicata (IV), vol. CXX p.P. 329-396 (1979).

[11] Da Prato, G., Grisvard, P.: maximal regularity for evolution equations by interpolation and extrapolation. To appear in Jouranl of Functional Analysis.

[12] Da Prato, G., Iannelli, M., Tubaro, L.: Some Results on Linear Stochastic Differential Equations in Hilbert Spaces, Stochastic, 6, 105-116 (1982).

[13] Da Prato, G., Iannelli, M., Tubaro, L.: On the path regularity of a Stochastic Process in a Hilbert space defined by the Itô integral, Stochastic, 6, 315-322 (1982).

[14] Da Prato, G., Iannelli, M., Tubaro, L.: Semi-linear stochastic differential equations in Hilbert spaces, Boll. U.M.I., 5, 168-185 (1979).

[15] Dawson, D.A.: Stochastic evolution equations, Math. Biosciences, 15, 287-316 (1972).

[16] Dunford, N., Schwartz, J.: Linear Operators II, Interscience (1963).

[17] Fleming, W.H.: Distribute parameter stochastic systems in population biology, Lec. Notes Econ. Math. Syst., 107, 179-191 (1975).

[18] Grisvard, P.: Commutativité de deux foncteurs d'interpolation et applications, J. Math. pures et appl., 45,

[19] Ichikawa, A.: Linear stochastic evolution equations in Hilbert spaces, J. Diff. Equat., 28, 266-283 (1978).

[20] Ichikawa, A.: Stability of semilinear evolution equations. J. Math. Analysis and Appl., 90, 13-44 (1982).

[21] Ichikawa, A.: Semilinear Stochastic Evolution Equations:

[22] Kato, T.: Fractional powers of dissipative operators II, J. Math. Soc. Japan, 14 (1962).

[23] Kotelenez, P.: A submartingale type inequality with applications to stochastic evolution equations, Stochastics 8

[24] Krylov,N.V., Rozovskii, B.L.: Itô equations in Banach spaces and strongly parabolic stochastic partial differential equations, Soviet Math. Dokl., 20, 1267-1271 (1979).

[25] Krylov, N.V., Rozovskii, B.L.: Stochastic evolution equations, Itogi Nauki i Tekhniki, Seyia Sovremennye Problemy Matenatiki, vol. 14, 71-146 (1979).

[26] Lions, J.L.: Espaces d'interpolation et domaines de puissances fractionnaires d'opérateurs, J. Math. Soc. Japan, 14, 234-241 (1962).

[27] Lions, J.L, Magenes, E.: Problemes aux limites non-homogenes et applications, Dunod, Paris, (1968).

[28] Lions, J.L., Peetre, J.: Sur une classe d'espaces d'interpolation, Publ. I.H.E.S., 19, 5-68 (1964).

[29] Metivier, M., Pistone, G.: Une formule d'isometrie pour l'integrale stochastique hilbertienne et équations d'évolution linéaires stochastiques. Z. Wahrscheinlichkeitstheorie verw. Gabiete, 33, 1-18 (1975).

[30] Pardoux, E.: Equations aux dérivés partielles stochastiques non linéaires monotones, Thèse, Université, Paris XI (1975).

[31] Pardoux, E.: Stochastic partial differential equations and filtering of diffusion processes, Stochastics, 3, (1979).

[32] Yosida, K.: Functional Analysis and Semigroups, Springer-Verlag (1965).

Egbert Dettweiler

STOCHASTIC INTEGRATION OF BANACH SPACE VALUED FUNCTIONS

1. INTRODUCTION

Let M be a square integrable (real valued, right continuous) martingale relative to a certain filtration. Then the starting point of stochastic integration theory is the L^2-isometry between the space of square integrable predictable functions X (relative to the Doléans measure of M^2) and the space of the stochastic integrals $\int X dM$. This result extends without major difficulties to the case that M is Hilbert space valued and X belongs to a suitable space of operator-valued predictable functions.

More essential difficulties arise in the case that the state space of the stochastic integrals is a general Banach space. It was proved in [1] that an inequality of the form

$$\mathbb{E}\|\int X dM\|^2 \leq C\int \|X\|^2 d\mu \quad \text{(with a constant } C>0\text{)}$$

is valid in general if and only if the state space E of the integral process is 2-smoothable - in the sense that E has an equivalent 2-uniformly smooth norm (see e.g.[6]).

Since the class of 2-smoothable spaces is a rather restricted class of Banach spaces (to give an impression: only for $2 \leq p < \infty$ the L^p-spaces belong to this class), it seems to be an interesting problem to study the possibility of stochastic integration also on those Banach spaces having a bad geometry - in the sense that they are not 2-smoothable. This is the aim of the present paper.

The problems arising with Banach space valued stochastic integrals are mainly connected with the geometry of the state space of the integral process. The state space of the integrator process is not so important in this context. Indeed, even

L. Arnold and P. Kotelenez (eds.), Stochastic Space-Time Models and Limit Theorems, 53–79.

if the integrator process is one-dimensional, the main difficulties are the same.

For this reason we mostly restrict ourselves to the case that the integrator process is the one-dimensional Brownian motion. All essential problems remain with this restriction. Moreover, it turns out that generalizations to more general integrator processes (e.g. Banach space valued martingales, infinite-dimensional Wiener process etc.) can be derived from our results.

In the first part of the paper we introduce a notion of the stochastic integral of Banach space valued functions relative to Brownian motion which works on every (real, separable) Banach space. This notion has only some theoretical meaning, since on a general Banach space there are no nice criteria for stochastic integrability - as the classical isometry or an inequality of the form indicated above. Nevertheless, it shows that the stochastic integral of elementary functions can always be extended to a larger (Banach) space of functions.

The second part is concerned with the integrability of special classes of functions and applications to stochastic integral equations on arbitrary Banach spaces. We study especially the problem of giving conditions that a Lipschitz function $g: \mathbb{R}_+ \times E \longrightarrow F$ (E,F Banach spaces) extends to a Lipschitz function on the space of E-valued square integrable, predictable processes X in the sense that

(1.) the integral $\int g(s,X_s)d\xi_s$ always exists ($\xi :=$ Brownian motion) and that

(2.) there exists a constant $C>0$ such that for any two processes X,Y as above

$$\mathbb{E}\|\int(g(s,X_s)-g(s,Y_s))d\xi_s\|^2 \leq C\,\mathbb{E}\int\|X_s-Y_s\|^2 ds.$$

Both properties of course hold for any classical Lipschitz function, if E and F are 2-smoothable, and it is well known from this case that property (2.) is intimately connected with the problem of proving existence and uniqueness of solutions of certain stochastic integral equations, where g is the integrand of the stochastic integral term.

In case that g does not depend on t there is a simple necessary and sufficient characterization that g has properties (1.) and (2.): g can be factorized in the form $g = T \circ f + x$, where f is a classical Lipschitz function from E to a Banach space G and T is a special bounded linear operator from G to F which will be called 2-smooth. Similar but more complicated characterizations hold in the case that g depends on t.

It follows from these results that the class of 2-smooth operators (already mentioned in [6], see also [7]) is an interesting class from the viewpoint of infinite-dimensional stochastic integration, and we present some examples of operators belonging to this class.

Finally, we also use the class of 2-smooth operators to define for every Banach space E a reasonable large space of E-valued predictable processes X, for which the existence of the stochastic integral $\int X d\xi$ follows as in the classical situation by an inequality of the form

$$\mathbb{E}\,\|\int X\, d\xi\|^2 \leq C\, \mathbb{E}\int \|f(X)\|^2 ds \ .$$

2. GENERAL DEFINITION OF THE STOCHASTIC INTEGRAL

Let $(\Omega, \mathcal{F}, P)$ be a fixed probability space and let $(\mathcal{F}_t)_{t \geq 0}$ be a <u>standard</u> <u>filtration</u> of Ω, i.e. we assume that $\mathcal{F}_0$ contains all P-null sets of the P-completion of $\mathcal{F}_\infty := \sigma(\mathcal{F}_t : t \geq 0)$ and that $(\mathcal{F}_t)$ is right continuous. Let $\xi = (\xi_t)$ denote a (one-dimensional) Brownian motion defined on Ω, and assume that $(\mathcal{F}_t)$ has the following additional properties: ξ is $(\mathcal{F}_t)$-adapted and $\xi_t - \xi_s$ is independentfrom $\mathcal{F}_r$ for $0 \leq r \leq s \leq t$.

Let $\mathcal{P}$ denote the σ-algebra of <u>predictable</u> subsets of $\mathbb{R}_+ \times \Omega$. By definition $\mathcal{P}$ is generated by the family $\mathcal{R}$ of the predictable rectangles $]s,t] \times F$ $(0 \leq s \leq t, F \in \mathcal{F}_s)$ or $\{0\} \times F$ $(F \in \mathcal{F}_0)$.

Now let E be a real, separable Banach space. We call a function $f: \mathbb{R}_+ \times \Omega \longrightarrow E$ an <u>elementary</u> (predictable E-valued) <u>function</u>, if f is of the form $f = \sum_{k=0}^{n-1} x_k 1_{]t_k, t_{k+1}] \times F_k}$ ($x_k \in E, 0 \leq t_0 < t_1 < \ldots < t_n$, $F_k \in \mathcal{F}_{t_k}$). By $\mathcal{E} := \mathcal{E}(E)$ we denote the vector space of all elementary functions. Let λ denote the Lebesgue measure on $\mathbb{R}_+$. If we identify functions in $\mathcal{E}$ which differ only on a $\lambda\otimes P$-null set, we can view $\mathcal{E}$ as a subspace of $L(E', L^2(\mathbb{R}_+ \times \Omega, \mathcal{P}, \lambda\otimes P))$ with its uniform operator norm, which we denote by $\|\cdot\|_{(2)}$. Let $\mathbb{L}_{(2)}$ denote the completion of $\mathcal{E}$ with respect to $\|\cdot\|_{(2)}$. Then the elements of $\mathbb{L}$ will be called (square integrable, predictable) <u>cylindrical</u> <u>functions</u>.

For $f \in \mathcal{E}$ of the above form the <u>stochastic integral</u> $\int f d\xi$ of f relative to ξ is defined as

$$\int f d\xi := \int f(t) d\xi_t := \sum_{k=0}^{n-1} x_k 1_{F_k} (\xi_{t_{k+1}} - \xi_{t_k}) .$$

Let $\|f\|_{2,\xi}$ denote the L^2-norm of $\int f d\xi$. Then the one-dimensional L^2-isometry implies $\|f\|_{2,\xi} \geq \|f\|_{(2)}$ and hence the completion $\mathbb{L}_\xi^2$ of $\mathcal{E}$ relative to the norm $\|\cdot\|_{2,\xi}$ is contained in $\mathbb{L}_{(2)}$. The elements of $\mathbb{L}_\xi^2$ will be called ξ-<u>square</u> <u>integrable</u> <u>cylindrical</u> <u>functions</u> and for $f \in \mathbb{L}_\xi^2$ the stochastic integral $\int f d\xi$ of f relative to ξ is defined as

$$\int f d\xi := \int f(t) d\xi_t := \lim_{n\to\infty} \int g_n d\xi ,$$

if (g_n) is a sequence in $\mathcal{E}$ with $\lim_{n\to\infty} \|f - g_n\|_{2,\xi} = 0$.

2.1 <u>Proposition</u>. If $f \in \mathbb{L}_\xi^2$, then also $f 1_{[0,t]} \in \mathbb{L}_\xi^2$ for all $t \geq 0$, and $(\int_0^t f(s) d\xi_s)_{t\geq 0}$ (where of course $\int_0^t f(s) d\xi_s := \int 1_{[0,t]} f d\xi$) can be viewed (and we will do so in the following) as a <u>continuous</u> square integrable martingale - in the sense that there exists a continuous version.

<u>Proof</u>: Let (g_n) be a sequence in $\mathcal{E}$ with $\lim \|f - g_n\|_{2,\xi} = 0$. For every $n \geq 1$ the process $(\int g_n 1_{[0,t]} d\xi)_{t\geq 0}$ is a continuous martingale. This implies

$$\mathbb{E} \| \int 1_{[0,t]} (g_n - g_m) d\xi \|^2 \leq \| g_n - g_m \|_{2,\xi}^2$$

for all $n, m \in \mathbb{N}$ and hence $f 1_{[0,t]} \in \mathbb{L}_\xi^2$. The martingale property of $(\int f 1_{[0,t]} d\xi)_{t\geq 0}$ follows from

$$\mathbb{E}[\int f d\xi | \mathcal{F}_t] = \lim_{n\to\infty} \mathbb{E}[\int g_n d\xi | \mathcal{F}_t] = \lim_{n\to\infty} \int g_n 1_{[0,t]} d\xi = \int_0^t f d\xi .$$

Now let $(\delta_k), (\varepsilon_k), (\eta_k)$ be sequences of positive numbers with the properties

$$\varepsilon_m = \sum_{k\geq m} \delta_k, \ \lim_{m\to\infty} \varepsilon_m = 0 \text{ and } \sum_{k\geq 1} \delta_k^{-2} \eta_k^2 < \infty ,$$

and choose a sequence (n_k) in $\mathbb{N}$ such that $\| g_{n_{k+1}} - g_{n_k} \|_{2,\xi} \leq \eta_k$. Then we get

$$P[\sup_{t\geq 0} \| \int_0^t (g_{n_{k+1}} - g_{n_k}) d\xi \| \geq \delta_k] \leq \delta_k^{-2} \eta_k^2$$

and hence by the Borel-Cantelli lemma

$$P[\exists m \forall l \geq k \geq m: \sup_{t\geq 0} \| \int_0^t (g_{n_l} - g_{n_k}) d\xi \| \leq \varepsilon_k]$$

$$\geq P[\exists m \forall k \geq m: \sup_{t\geq 0} \| \int_0^t (g_{n_{k+1}} - g_{n_k}) d\xi \| \leq \delta_k] = 1 .$$

This proves that there exists a continuous version for the martingale $(\int_0^t f d\xi)_{t\geq 0}$. ┘

Now we extend the stochastic integral to spaces of cylindrical functions which are only locally integrable in the following sense:

A cylindrical function $f \in L(E', L^0(\mathbb{R}_+ \times \Omega, \mathcal{P}, \lambda \otimes P))$ is said to be <u>locally ξ-integrable</u>, if there exists an increasing sequence (T_n) of stopping times with $\lim_{n\to\infty} T_n = \infty$ a.s. such that

(i) $f 1_{[0,T_n]} \in \mathbb{L}_\xi^2$ for all $n \geq 1$ and

(ii) $\int_0^t f d\xi := \lim_{n\to\infty} \int f 1_{[0, t\wedge T_n]} d\xi$ exists a.s. for all $t \geq 0$.

$\mathbb{L}_\xi^{loc}$ will denote the space of all locally ξ-integrable cyl. functions. It is not difficult to see that there exists always a continuous version of $(\int_0^t f d\xi)_{t\geq 0}$ and in this sense $(\int_0^t f d\xi)_{t\geq 0}$ is a continuous local martingale.

2.2 <u>Proposition</u>. For every $f \in \mathbb{L}_\xi^{loc}$ there exist a sequence (g_n) in $\mathcal{E}$ and a double-sequence $(T_{n,k})$ of stopping times increasing in n and k and having the property $\lim_{k\to\infty} \sup_{n\geq 1} T_{n,k} = \infty$ a.s. such that

(1) for all $k \geq 1$, $(\int g_n 1_{[0,T_{n,k}]} d\xi)_{n \geq 1}$ is a Cauchy sequence in $\mathbb{L}^2_\xi$,

(2) for all $t \geq 0$, $(\int_0^t g_n d\xi)_{n \geq 1}$ is a Cauchy sequence in probability, and

(3) $f 1_{[0,T_k]} = \lim_{n\to\infty} g_n 1_{[0,T_{n,k}]}$ in $\mathbb{L}_{(2)}$, where $T_k = \sup_{n \geq 1} T_{n,k}$.

Conversely, suppose that (g_n) is a sequence in $\mathcal{E}$ and $(T_{n,k})$ is a double-sequence of stopping times as above such that (1) and (2) hold. Then (3) defines (!) an $f \in \mathbb{L}^{loc}_\xi$.

<u>Proof</u>: I. Let $f \in \mathbb{L}^{loc}_\xi$ and let (T_n) be a localizing sequence of stopping times for f such that conditions (i) and (ii) hold. Then for every fixed $n \geq 1$ there exists a sequence $(f_{n,k})_{k\geq 1}$ in $\mathcal{E}$ such that $\lim_{k\to\infty} f_{n,k} = f 1_{[0,T_n]}$ in $\mathbb{L}^2_\xi$. Now let (ε_n) be a given null-sequence of positive numbers and put $t_n = \varepsilon_n^{-1}$ for all $n \geq 1$. Then we can find for every $n \geq 1$ an m_n such that $P[T_{m_n} \leq t_n] \leq \varepsilon_n$ and then a k_n such that $\| f_{m_n,k_n} 1_{[0,T_{m_n}]} - f 1_{[0,T_{m_n}]} \|^2_{2,\xi} \leq \varepsilon_n^3$. For every $t \leq t_n$ we get for the sequence (g_n) given by $g_n := f_{m_n,k_n}$:

$$\mathbb{E}\Big[\inf\Big(1, \sup_{0\leq s\leq t} \| \int g_n 1_{[0,s]} d\xi - \int f 1_{[0,s]} d\xi \|\Big)\Big]$$
$$\leq P[T_{m_n} \leq t_n] + \int 1_{[T_{m_n} > t_n]} \inf\Big(1, \sup_{s\leq t} \| \int_0^s g_n d\xi - \int_0^s f d\xi \|\Big) dP$$
$$\leq \varepsilon_n + \mathbb{E}\Big[\inf\Big(1, \sup_{s\leq t} \| \int g_n 1_{[0,s\wedge T_{m_n}]} d\xi - \int f 1_{[0,s\wedge T_{m_n}]} d\xi \|\Big)\Big]$$
$$\leq 2\varepsilon_n + P\Big[\sup_{s\leq t} \| \int g_n 1_{[0,s\wedge T_{m_n}]} d\xi - \int f 1_{[0,s\wedge T_{m_n}]} d\xi \| \geq \varepsilon_n\Big]$$
$$\leq 2\varepsilon_n + \varepsilon_n^{-2} \mathbb{E} \| \int g_n 1_{[0,T_{m_n}]} d\xi - \int f 1_{[0,T_{m_n}]} d\xi \|^2 \leq 3\varepsilon_n ,$$

where the last line follows from Doob's inequality. This proves that $(\int_0^t g_n d\xi)_{n \geq 1}$ is a Cauchy sequence relative to the Fréchet norm $\|.\|_0$ on $L^0(\Omega, \mathcal{F}, P; E)$ defined by $\|X\|_0 := \mathbb{E} \inf(1, \|X\|)$, i.e. we have proved that (2) holds. To prove (1) and (3) we define for any $g \in \mathbb{L}^{loc}_\xi$ and every $c \geq 0$ the following stopping times:

$$T_c(g) := \inf\{t\geq 0 : \|\int_0^t g d\xi\| > c\} \text{ and}$$
$$T_{c-}(g) := \inf\{t\geq 0 : \|\int_0^t g d\xi\| \geq c\}.$$

Then it follows from the continuity of the process $(\int f 1_{[0,t]} d\xi)_{t\geq 0}$ that for every $c\geq 0$

$$(*)\ T_{c-}(f) \leq \underline{\lim}\ T_{c-}(g_n) \leq \overline{\lim}\ T_c(g_n) \leq T_c(f).$$

We put $T_{n,k} := T_{k-}(f) \wedge \inf_{m\geq n} T_k(g_m)$ for every $n\geq 1, k\geq 1$. Then for every fixed k the sequence $(\int_0^{T_{n,k}} g_n d\xi)_{n\geq 1}$ is uniformly bouded by k and it follows from $(*)$ and (2) that (1) and (3) hold.

II. If (g_n) is a sequence in $\mathcal{E}$ such that for a double-sequence $(T_{n,k})$ as above conditions (1) and (2) hold, then for every $k\geq 1$ we put

$$f_k := \lim_{n\to\infty} g_n 1_{[0,T_{n,k}]} \quad (\text{in } \mathbb{L}_{(2)}).$$

It follows from (1) that

$$f_j 1_{[0,T_k]} = f_k \quad \text{for } k < j \ (\text{where } T_k := \sup_{n\geq 1} T_{n,k}),$$

hence the cylindrical function f defined by f_k on $[0,T_k]$ is well defined, and it is not difficult to see that f has properties (i) and (ii), i.e. f is locally ξ-integrable.|

<u>Remark</u>: $\mathbb{L}_\xi^{loc}$ can also be obtained (as $\mathbb{L}_\xi^2$) as a completion of $\mathcal{E}$ relative to a suitable Fréchet norm. For every $g\in\mathcal{E}$ we put

$$\|g\|_{0,\xi} := \sup_{t\geq 0} t^{-1}\, \mathbb{E}[\inf(1, \sup_{0\leq s\leq t}\|\int_0^s g d\xi\|)].$$

Then $\mathbb{L}_\xi^{loc} = (\mathcal{E}, \|.\|_{0,\xi})^\sim$.

Propositions 2.1 and 2.2 show that it is possible to define on an arbitrary Banach space E the stochastic integral for a space of (cylindrical) functions which is always larger than $\mathcal{E}$ (since $\mathbb{L}_\xi^2$ is a Banach space). There are two problems hidden behind our definition of the stochastic integral. The minor problem is that $\mathbb{L}_\xi^2$ need not be a space of E-valued <u>functions</u>. This could be avoided for example in the following way. Consider instead of $\mathbb{L}_\xi^2$ the smaller Banach space

$$\mathbb{L}_\xi^{(2)} := \mathbb{L}_\xi^2 \cap L^2(\mathbb{R}_+\times\Omega, \mathcal{P}, \lambda\otimes P; E)$$

(with the norm $\|\cdot\|_{(2),\xi} := \max(\|\cdot\|_{2,\xi}, \|\cdot\|_2)$) which is again a Banach space containing $\mathcal{E}$.

The major problem consists in the lack of a general criterion to decide whether a given function belongs to $\mathbb{L}^2_\xi$, i.e. can be approximated by a sequence of elementary functions in the $\mathbb{L}^2_\xi$-norm. In the case of a Hilbert space this criterion is of course given by the isometry between $\mathbb{L}^2_\xi$ and $L^2(\lambda\otimes P, E)$. If E is not a Hilbert space, then this isometry is no longer valid. More precisely, we have the following result (which is proved in [1]):

2.3 Theorem.

(1) For a Banach space E the inclusion $\mathbb{L}^2_\xi \supset L^2(\lambda\otimes P, E)$ holds if and only if E is 2-smoothable (i.e. there exists an equivalent norm $|\cdot|$ on E such that $(E, |\cdot|)$ is 2-uniformly smooth, see [6] or [7]).
(2) The inclusion $\mathbb{L}^2_\xi \subset L^2(\lambda\otimes P, E)$ holds if and only if E is 2-convexifiable (i.e. there exists an equivalent norm $|\cdot|$ on E such that $(E, |\cdot|)$ is 2-uniformly convex).
(3) $\mathbb{L}^2_\xi = L^2(\lambda\otimes P, E)$ if and only if E is isomorphic to a Hilbert space.

The theorem indicates that for not 2-smoothable spaces it will be a problem to give general criteria that given functions belong to $\mathbb{L}^2_\xi$. This will be the subject of the following paragraphs.

3. LIPSCHITZ FUNCTIONS ON BANACH SPACES

Let F be a second Banach space. Then $\mathrm{Lip}(\mathbb{R}_+\times E, F)$ will denote the space of all Borel measurable functions $g: \mathbb{R}_+\times E \to F$ with the property

(Lip) $\sup_{t\geq 0}\|g(t,o)\| < \infty$ and there exists a constant $c \geq 0$ such that for all $x, y \in E$

$$\sup_{t\geq 0} \|g(t,x) - g(t,y)\| \leq c\|x-y\|.$$

The subspace of Lipschitz functions which do not depend on t will be denoted by $\mathrm{Lip}(E,F)$. For every $g \in \mathrm{Lip}(\mathbb{R}_+\times E, F)$ let $L(g)$ be the smallest constant $c \geq \sup\|g(t,o)\|$ such that (Lip) holds and suppose for a moment that F is 2-smoothable. Then it is known (see [1]) that there exists a constant $C \geq 0$

such that for all $g \in \mathrm{Lip}(\mathbb{R}_+ \times E, F)$ and all $X, Y \in L^2(\lambda \otimes P, E)$

$$\mathbb{E}\|\int g(t,X_t)d\xi_t - \int g(t,Y_t)d\xi_t\|^2$$
$$\leq C^2\, \mathbb{E}\int \|g(t,X_t) - g(t,Y_t)\|^2 dt$$
$$\leq C^2 L(g)^2\, \mathbb{E}\int \|X_t - Y_t\|^2\, dt\,.$$

In the notation of section 2 this means

$$(*)\quad \|g(.,X.) - g(.,Y.)\|_{2,\xi} \leq CL(g)\,\|X - Y\|_2\,,$$

i.e. $g \in \mathrm{Lip}(L^2(\lambda \otimes P, E), \mathbb{L}^2_\xi(F))$.

It is well known that inequality $(*)$ is the essential inequality in proving existence and uniqueness for the solutions of diffusion equations, where g is the function occuring in the diffusion term (see [3], [4], [5] and in the Banach space case e.g. [2]). If F is not necessarily 2-smoothable, then in general $(*)$ is not valid because of the following two problems:

<u>Problem 1</u>: If $g \in \mathrm{Lip}(\mathbb{R}_+ \times E, F)$ and $X \in L^2(\lambda \otimes P, E)$, then not necessarily $g(X) := g(.,X.) \in \mathbb{L}^2_\xi$, i.e. the stochastic integral $\int g(X)d\xi$ need not exist.

<u>Problem</u> 2: Even if $g(X) \in \mathbb{L}^2_\xi$, then inequality $(*)$ need not be valid.

For these reasons we will now study a reasonable large subclass of Lipschitz functions, for which both problems have a positive answer.

We introduce the following notational conventions. Let Π denote the family of all finite sequences $\Delta = (t_k)_{0 \leq k \leq n}$ in $\mathbb{R}_+$ with $t_k < t_{k+1}$ for $0 \leq k \leq n-1$. Π_t will denote the subfamily of all $\Delta \in \Pi$ with $t_0 \geq t$. For every $\Delta \in \Pi$ we put

$$\varphi^\Delta := \sum_{k=0}^{n-1} t_k 1_{]t_k, t_{k+1}]}\,,$$

and for any function f defined on $\mathbb{R}_+$ we denote by f^Δ the function $f \circ \varphi^\Delta$.

Let us call a function $f: \mathbb{R}_+ \longrightarrow E$ <u>ξ-continuous</u>, if the following two conditions hold:

(i) $f \in \mathbb{L}^2_\xi(E)$ and $\lim_{|\Delta| \to 0} \| f^\Delta - f \|_{2,\xi} = 0$ (where as usual $|\Delta| := \max(t_{k+1} - t_k)$ for $\Delta = (t_k)_{0 \le k \le n}$).

(ii) $\lim_{h \downarrow 0} h^{-1}\, \mathbb{E} \| \int_t^{t+h} (f(s) - f(t))\, d\xi_s \|^2 = 0$ for all $t \geq 0$.

By $C_\xi(\mathbb{R}_+, E)$ we will denote the space of all ξ-continuous E-valued functions on $\mathbb{R}_+$.

Now we will say that a Borel function $g: \mathbb{R}_+ \times E \longrightarrow F$ is a ξ-<u>Lipschitz</u> <u>function</u>, if g has the following properties:

(1) $g(.,x) \in C_\xi(\mathbb{R}_+, F)$ for all $x \in E$, and

(2) there exists a constant $C \geq 0$ such that for all $X, Y \in \mathcal{E}(E)$ and all $Z \in \mathcal{E}(\mathbb{R})$

$$\mathbb{E} \| \int (g(X) - g(Y))\, Z\, d\xi \|^2 \leq C^2\, \mathbb{E} \int (Z \| X - Y \|)^2 d\lambda .$$

We denote by $C(g)$ the smallest constant C for which inequality (2) holds, and we will write $\mathrm{Lip}_\xi(\mathbb{R}_+ \times E, F)$ for the space of all ξ-Lipschitz functions. Finally, $\mathrm{Lip}_\xi(E,F)$ will denote the subspace of all ξ-Lipschitz functions, which do not depend on t (so that condition (1) is trivially fulfilled).

We will now look for simpler characterizations of the space $\mathrm{Lip}_\xi(\mathbb{R}_+ \times E, F)$. First we prove a renorming lemma, which will be essential for the later characterizations (cf.[6], theorem 3.1 for the origin of the main trick in the proof).

3.2 <u>Lemma</u>. For every $g \in \mathrm{Lip}_\xi(\mathbb{R}_+ \times E, F)$ there exists a family $(n_t)_{t \geq 0}$ of norms on F with the following properties:

(i) $\| x \| \leq n_t(x) \leq n_s(x) \leq \sqrt{2} \| x \|$ for all $x \in F$ and $0 \leq s \leq t$, and

$\lim_{s \downarrow t} n_s(x) = n_t(x)$ for all $x \in F$, $t \geq 0$.

(ii) There exists a constant $D = D(g)$ such that for a given $N(0,1)$-distributed Gaussian random variable ξ, for all $t \geq 0$, $x \in F$ and $y, z \in E$

$$\mathbb{E}\, n_t(x + (g(t,y) - g(t,z))\, \xi)^2 \leq n_t(x)^2 + D^2 \| y - z \|^2 .$$

Proof: We put $\mathcal{F}_{s,t} := \sigma(\xi_r - \xi_s;\ s \leq r \leq t)$ and denote by $\mathcal{P}_s$ the -algebra of predictable subsets of $[s,\infty]\times\Omega$ defined by the filtration $(\mathcal{F}_{s,t})_{t\geq s}$. We will use the notation $X \in \mathcal{E}_s(E)$ to express that $X \in \mathcal{E}(E)$ is $\mathcal{P}_s$-predictable.

Now we define for every $t \geq 0$, $x \in F$

$$n_t(x)^2 := \sup_{X,Y,Z} \mathbb{E}\|x + \int_t^\infty Z(g(X)-g(Y))d\xi\|^2 - 2C(g)^2 \int_t^\infty \mathbb{E}(Z\|X-Y\|)^2 d\lambda ,$$

where the supremum is taken over all $X,Y \in \mathcal{E}_t(E)$ and $Z \in \mathcal{E}_t(\mathbb{R})$. We will prove that the family (n_t) has properties (i) and (ii).

Let us first show that property (ii) holds. We put $G := g(.,y)-g(.,z)$ and prove that for every fixed $\varepsilon > 0$ the following inequality holds:

$$(1)\quad \mathbb{E}\, n_t(x+\int_s^t G d\xi)^2 \leq n_s(x)^2 + C(t-s)\|y-z\|^2 + \varepsilon ,$$

where $t > s$ and $C = 2\, C(g)^2$. Then the inequality in (ii) follows from the ξ-continuity of g and the continuity property (i) of the family (n_t), if (i) is proved.

We introduce the abbreviations

$$N_t^1(x,X,Y,Z) := \mathbb{E}\|x + \int_t^\infty Z(g(X)-g(Y))d\xi\|^2 ,$$

$$N_t^2(x,X,Y,Z) := C \int_t^\infty \mathbb{E}(Z\|X-Y\|)^2 d\lambda \quad \text{and}$$

$$N_t(x,X,Y,Z) := N_t^1(x,X,Y,Z) - N_t^2(x,X,Y,Z)$$

for $x \in F$, $X,Y \in \mathcal{E}_t(E)$ and $Z \in \mathcal{E}_t(\mathbb{R})$. W.l.o.g. we may and do assume that $N_t^2(x,Y,Y,Z) \leq k$ for a given constant $k \geq 0$. The general inequality (1) then follows by a limit argument.

Now let for a given $\delta > 0$ (the dependence of ε will be specified later) $(V_j)_{j\geq 1}$ denote a partition of F into Borel sets such that for every $j \geq 1$ and all $u,v \in V_j$:

$$\|u-v\| < \delta \quad \text{and} \quad |n_t(x+u)^2 - n_t(x+v)^2| < \delta .$$

The existence of such a partition follows from the separability of F and the integrability of $n_t(x+\eta)$, where $\eta := \int_s^t G d\xi$.

We choose a fixed $u_j \in V_j$ for every $j \geq 1$. By definition of n_t there exist for every $j \geq 1$ a pair $X_j, Y_j \in \mathcal{E}_t(E)$ and a $Z \in \mathcal{E}_t(\mathbb{R})$ such that

$$n_t(x+u_j)^2 \leq N_t(x+u_j, X_j, Y_j, Z_j)^2 + \delta$$

for every $j \geq 1$. Now we define

$$X := y\, 1_{]s,t]\times\Omega} + \sum_{j\geq 1} X_j\, 1_{[\eta\in V_j]},$$
$$Y := z\, 1_{]s,t]\times\Omega} + \sum_{j\geq 1} Y_j\, 1_{[\eta\in V_j]} \quad \text{and}$$
$$Z := 1_{]s,t]\times\Omega} + \sum_{j\geq 1} Z_j\, 1_{[\eta\in V_j]}.$$

Then we have $X, Y \in \mathcal{E}_s(E)$, $Z \in \mathcal{E}_s(\mathbb{R})$ and

$$\eta + \int_t^\infty Z(g(X)-g(Y))d\xi = \int_s^\infty Z(g(X)-g(Y))d\xi$$

It follows from the ξ-Lipschitz property of g and the choice of the partition (V_j) that there is a constant $K > 0$ such that

$$(2) \quad \Big| \mathbb{E}\|x + \eta + \int_t^\infty Z(g(X)-g(Y))d\xi\|^2 - \mathbb{E}\|x + \zeta + \int_t^\infty Z(g(X)-g(Y))d\xi\|^2 \Big| < K\delta$$

where $\zeta := \sum_{j\geq 1} u_j 1_{[\eta\in V_j]}$. With the aid of (2) we finally get the following chain of inequalities, which will prove inequality (1):

$$\mathbb{E}\, n_t(x + \int_s^t G d\xi)^2$$
$$\leq \sum_{j\geq 1} P[\eta \in V_j]\, n_t(x + u_j)^2 + \delta$$
$$\leq \sum_{j\geq 1} P[\eta \in V_j]\, N_t(x+u_j, X_j, Y_j, Z_j) + 2\delta$$
$$= \mathbb{E}\|x + \zeta + \int_t^\infty Z(g(X)-g(Y))d\xi\|^2 - C\int_t^\infty \mathbb{E}(Z\|X-Y\|)^2 d\lambda + 2\delta$$
$$\leq \mathbb{E}\|x+\eta+\int_s^\infty Z(g(X)-g(Y))d\xi\|^2 - C\int_t^\infty \mathbb{E}(Z\|X-Y\|)^2 d\lambda + \delta(2+K)$$

$$\leq \mathbb{E}\|x+ \int_s^\infty Z(g(X)-g(Y))d\xi\|^2$$
$$- C\int_t^\infty \mathbb{E}(Z\|X-Y\|)^2 d\lambda + \delta(2+K)$$
$$= \mathbb{E}\|x + \int_s^\infty Z(g(X)-g(Y))d\xi\|^2$$
$$- C\int_s^\infty \mathbb{E}(Z\|X-Y\|)^2 d\lambda + C(t-s)\|y-z\|^2 + \delta(2+K)$$
$$\leq n_s(x)^2 + C(t-s)\|y-z\|^2 + \varepsilon \text{, if } (2+K)\delta \leq \varepsilon.$$

Now let us prove (i). That (n_t) is a family of norms follows easily from the definition: every n_t is homogeneous and also convex. Putting X = Y in the definition of n_t as a supremum we get $\|x\| \leq n_t(x)$, and $n_t(x) \leq \sqrt{2}\,\|x\|$ follows from

$$\mathbb{E}\|x + \int_t^\infty Z(g(X)-g(Y))d\xi\|^2$$
$$\leq 2\|x\|^2 + 2C(g)^2 \int_t^\infty \mathbb{E}(Z\|X-Y\|)^2 d\lambda .$$

The continuity property can be obtained as follows: For $s \geq 0$, $x \in F$ and every $\varepsilon > 0$ there is a triple X, Y, Z as above such that

$$n_s(x)^2 \leq \mathbb{E}\|x + \int_s^\infty Z(g(X)-g(Y))d\xi\|^2$$
$$- C\int_s^\infty \mathbb{E}(Z\|X-Y\|)^2 d\lambda + \varepsilon$$
$$= \lim_{t\downarrow s}\Big[\mathbb{E}\|x + \int_t^\infty Z(g(X)-g(Y))d\xi\|^2$$
$$- C\int_t^\infty \mathbb{E}(Z\|X-Y\|)^2 d\lambda\Big] + \varepsilon$$
$$\leq \lim_{t\downarrow s} n_t(x)^2 + \varepsilon \leq n_s(x)^2 + \varepsilon . \qquad \rfloor$$

3.3 Remarks. (1) If g does not depend on t, then it is not necessary to consider the whole family (n_t) of norms. We then take the norm $n := n_0$.

(2) The inequality in 3.2(ii) can be interpreted as a classical Lipschitz property of g. Assume for simplicity $g \in Lip_\xi(E,F)$ and define for every $z \in F$:

$$m(z) := \sup_{y\in F} \left(\mathbb{E}\, n(y+z\xi)^2 - n(y)^2\right)^{1/2} .$$

(where ξ is N(o,1)-distributed). Then m is a norm on the subspace $G := \{z \in F : m(z) < \infty\}$ and inequality 3.2(ii) can be written as

$$m(g(x)-g(y)) \leq C(g)\|x-y\| \quad (x,y \in E).$$

We will use this fact later on.

3.4 Theorem. The following assertions are equivalent

(1) g is a ξ-Lipschitz function.

(2) There exists a family (n_t) of equivalent norms on F with properties (i) and (ii) of lemma 3.2.

(3) There exists a family (n_t) of equivalent norms on F with property (i) of lemma 3.2 and the following property (ii') instead of (ii):

(ii') there is a constant $D \geq o$ such that for a given Bernoulli variable ε, all $c \geq o$, $t \geq o$, $y \in F$, $x_1, x_2 \in E$

$$\mathbb{E}\, n_t(y + c(g(t,x_1)-g(t,x_2))\varepsilon)^2 \leq n_t(y)^2 + Dc^2\|x_1-x_2\|^2.$$

Proof: The implication (1) $\Rightarrow$ (2) is given by lemma 3.2. The equivalence of (2) and (3) is a consequence of the following observation: Let x,y be elements of a Banach space $(G, \|.\|)$ and $s \geq o$. Then the inequalities

(I_1) there exists a constant $C \geq o$ (independent of x,y and s) such that for all $c \geq o$

$$\mathbb{E}\|x + cy\xi\|^2 \leq \|x\|^2 + Cc^2 s, \text{ and}$$

(I_2) there exists a constant $D \geq o$ (also independent of x,y and s) such that for all $c \geq o$

$$\mathbb{E}\|x + cy\varepsilon\|^2 \leq \|x\|^2 + Dc^2 s$$

are equivalent: Suppose that (I_2) holds. Then

$$\begin{aligned} &\mathbb{E}\|x + cy\xi\|^2 + \|x\|^2 \\ &= \mathbb{E}\|x + cy\xi\|^2 + \|x - \mathbb{E}(cy\xi)\|^2 \\ &\leq \mathbb{E}(\|x+cy\xi\|^2 + \|x-cy\xi\|^2) \\ &\leq \mathbb{E}(2\|x\|^2 + 2Dc^2\xi^2 s) = 2\|x\|^2 + 2Dc^2 s \end{aligned}$$

and (I_1) holds with C = 2D. The proof of $(I_1)\Rightarrow(I_2)$ is similar and is essentially a consequence of the symmetry of ξ.

It remains to prove (2) $\Rightarrow$ (1). We first make the following observation. Let $0 \leq s \leq t$ be given, and suppose that

$$X_1, X_2 \in L^2(\Omega, \mathcal{F}_s, P; E), Y \in L^2(\Omega, \mathcal{F}_s, P; F), Z \in L^\infty(\Omega, \mathcal{F}_s, P).$$

Then it follows from (2) (see [7], proof of theorem 2.1) that the following inequality holds:

$$\mathbb{E}\left[n_t(Y+Z(g(s,X_1)-g(s,X_2))(\xi_t-\xi_s))^2 \mid \mathcal{F}_s\right] \leq n_s(Y)^2 + CZ^2\|X_1-X_2\|^2(t-s) \quad . \tag{I}$$

Now let $X, Y \in \mathcal{E}(E)$, $Z \in \mathcal{E}(\mathbb{R})$ be given and take an arbitrary $\Delta = (t_k)_{0 \leq k \leq n+1} \in \Pi$ such that for every $\omega \in \Omega$ the functions $X_\cdot(\omega)$, $Y_\cdot(\omega)$ and $Z_\cdot(\omega)$ are constant on the intervals $]t_k, t_{k+1}]$ $(0 \leq k \leq n)$ and vanish outside $[t_0, t_{n+1}]$. Then we obtain with the aid of (I) the inequality

$$\begin{aligned}
&\mathbb{E}\left\| \int Z(g^\Delta(X)-g^\Delta(Y))d\xi \right\|^2 \\
&= \mathbb{E}\left\| \sum_{k=0}^n Z_{t_k}(g(t_k,X_{t_k})-g(t_k,Y_{t_k}))(\xi_{t_{k+1}}-\xi_{t_k}) \right\|^2 \\
&\leq \mathbb{E}\left(\mathbb{E}\left[n_t\left(\sum_{k=0}^n Z_{t_k}(g(t_k,X_{t_k})-g(t_k,Y_{t_k}))(\xi_{t_{k+1}}-\xi_{t_k})\right)^2 \Big| \mathcal{F}_{t_n}\right]\right) \\
&\leq \mathbb{E}\left(n_{t_n}\left(\sum_{k=0}^{n-1} Z_{t_k}(g(t_k,X_{t_k})-g(t_k,Y_{t_k}))(\xi_{t_{k+1}}-\xi_{t_k})\right)^2\right) \\
&\qquad + C\,\mathbb{E}(Z_{t_n}\|X_{t_n}-Y_{t_n}\|)^2(t_{n+1}-t_n) \\
&\leq \ldots \leq C\sum_{k=0}^n \mathbb{E}(Z_{t_k}\|X_{t_k}-Y_{t_k}\|)^2(t_{k+1}-t_k) \\
&= C\int \mathbb{E}(Z_t\|X_t - Y_t\|)^2 dt \quad ,
\end{aligned}$$

and we get $g \in \mathrm{Lip}_\xi(\mathbb{R}_+ \times E, F)$ by the ξ-continuity of g.┘

3.5 <u>Corollary</u>. For $T \in L(E,F)$ the following assertions are equivalent:

(1) There exists a constant $C \geq 0$ such that for all $X \in L^2(\lambda \otimes P, E)$

$$\mathbb{E}\,\|\int TX\, d\xi\,\|^2 \leq C^2 \int \mathbb{E}\,\|X\|^2\, dt$$

(2) Let (ε_n) denote a fixed Bernoulli sequence and put $\mathcal{G}_n := \sigma(\varepsilon_1, \ldots, \varepsilon_n)$. Then there exists a constant $D \geq 0$ such that for every sequence (X_n) with $X_n \in L^2(\Omega, \mathcal{G}_n, P; E)$ for all $n \geq 1$:

$$\sup_{n \geq 1} \mathbb{E}\,\|\sum_{k=0}^{n} TX_k \varepsilon_{k+1}\|^2 \leq D^2 \sum_{k=0}^{n} \mathbb{E}\,\|X_k\|^2 .$$

(3) There exist a Banach space G and operators $S \in L(E,G)$, $R \in L(G,F)$ with $T = R \circ S$, where S has the additional property: there is a constant $K \geq 0$ such that for all $x \in E$ and all $y \in G$

$$\mathbb{E}\,\|y + Sx\varepsilon\|^2 \leq \|y\|^2 + K^2 \|x\|^2 .$$

Proof. We only have to observe that (1) means $T \in \mathrm{Lip}_\xi(E,F)$. Then the equivalence of (1), (2) and (3) follows from the theorem.

3.6 Definition. (cf. [6] and [7]) An operator $T \in L(E,F)$ is called 2-smooth, if one of the equivalent conditions of corollary 3.5 holds. The operator S in condition (3) of the corollary is called 2-uniformly smooth. $S^2(E,F)$ will denote the space of 2-smooth operators. For $T \in S^2(E,F)$ we denote by $\sigma_2(T)$ the smallest constant $D \geq 0$, for which the inequality in condition (2) of corollary 3.5 holds. Then $S^2(E,F)$ is a Banach space under the norm $\sigma_2(.)$. This Banach space will be investigated in more detail in the next paragraph.

3.7 Theorem. The following assertions are equivalent

(1) $g \in \mathrm{Lip}_\xi(\mathbb{R}_+ \times E, F)$.

(2) For every $t \geq 0$ there exist Banach spaces F_t, G_t, Lipschitz functions $h_t \in \mathrm{Lip}(E, G_t)$, operators $T_t \in L(G_t, F_t)$ which are 2-uniformly smooth, operators $S_t \in L(F_t, F)$ and for $s \leq t$ operators $R_{t,s} \in L(F_s, F_t)$ such that

(i) $\sup_{t \geq 0} \sigma_2(T_t) < \infty$, $\sup_{t \geq 0} \|S_t\| < \infty$, $\sup_{s \leq t} \|R_{t,s}\| \leq 1$ and

(ii) the following diagram commutes $(0 \leq s \leq t)$:

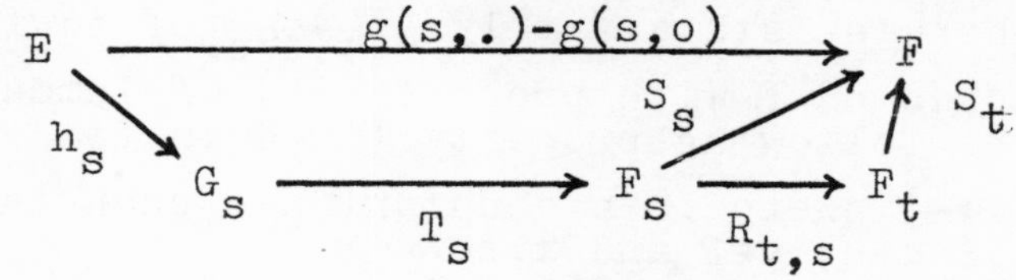

(iii) $g(.,x) \in C_\xi(\mathbb{R}_+,F)$ for all $x \in E$.

Proof: The implication (2) $\Longrightarrow$ (1) is proved exactly in the same way as the implication (2) $\Longrightarrow$ (1) of theorem 3.4. So we only prove (1) $\Longrightarrow$ (2). With the notations of theorem 3.4 we put

$$F_t := (F,n_t) \text{ , and } S_t := R_{t,s} := \text{Identity on } F \text{ .}$$

For every $z \in F$ we define

$$m_t(z) := \sup_{y \in F} \left[\mathbb{E}\, n_t(y + z\varepsilon)^2 - n_t(y)^2\right]^{1/2}$$

and put $G_t := \{ z \in F : m_t(z) < \infty \}$. Then G_t is a Banach space under the norm m_t, and the injection T_t from G_t into $F_t = F$ is obviously 2-uniformly smooth by the definition of m_t. Finally, it follows from theorem 3.4 that the Lipschitz function $h_t := g(t,.)-g(t,o)$ takes values in G_t, and the theorem is proved.

If the function g does not depend on t, the theorem gives a simple criterion that g is a ξ-Lipschitz function:

3.8 Corollary. A function $g: E \longrightarrow F$ belongs to $Lip_\xi(E,F)$ if and only if g is of the form $g = T \circ h + z$ where z is a fixed element in F, h is a "normal" Lipschitz function from E into a third Banach space G, and T is a 2-smooth operator from G to F.

In the case that g is differentiable the property $g \in Lip_\xi(\mathbb{R}_+ \times E,F)$ can also be characterized by an analogous property of the derivative g':

3.9 Proposition. Suppose that g is ξ-continuous and continuously differentiable in the second component and denote by g' the derivative $g' : \mathbb{R}_+ \times E \longrightarrow L(E,F)$. Then the following assertions are equivalent:

(1) $g \in \mathrm{Lip}_{\xi}(\mathbb{R}_+ \times E, F)$.

(2) There exists a family $(n_t)_{t \geq 0}$ of equivalent norms on F with property (i) of lemma 3.2 and the following property (ii-) instead of (ii):

(ii-) there is a constant $K \geq 0$ such that for all $t \geq 0$, $y \in F$ and $x, z \in E$

$$\mathbb{E}\, n_t(y + g'(t,x)z\,\varepsilon)^2 \leq n_t(y)^2 + K\|z\|^2 .$$

(3) For every $\Delta = (t_k) \in \Pi$ and all $X, Y \in L^2(\lambda \otimes P, E)$

$$\mathbb{E}\,\|\sum_k g'(t_k, X_{t_k}) Y_{t_k} (\xi_{t_{k+1}} - \xi_{t_k})\|^2 \leq C \sum_k \mathbb{E}\|Y_{t_k}\|^2 (t_{k+1} - t_k)$$

Proof: To prove (1) $\Rightarrow$ (2), let (n_t) denote the family of norms which is given by the equivalence (1) $\Rightarrow$ (3) of theorem 3.4. The inequality in (ii') implies that for every $t \geq 0$, $h > 0$, $y \in F$ and $x, z \in E$:

$$\mathbb{E}\, n_t(y + h^{-1}[g(t,x) - g(t,x+hz)]\,\varepsilon)^2 \leq n_t(y)^2 + D\|z\|^2 .$$

Since by assumption g is continuously differentiable we obtain (ii-) with $K := D$ for $h \to 0$.

The implication (2) $\Rightarrow$ (3) is proved exactly in the same way as (3) $\Rightarrow$ (1) of theorem 3.4. So it remains to prove (3) $\Rightarrow$ (1).

Let $\Delta = (t_k)_{1 \leq k \leq n}$ be fixed and $X^1, X^2 \in L^2(\lambda \otimes P, E)$, $Z \in L^\infty(\lambda \otimes P)$ be given. We put $X = X^1$ and $Y = Z(X^2 - X^1)$ and write ξ_k for the increment $\xi_{t_{k+1}} - \xi_{t_k}$. Then the following inequalities hold:

$$\mathbb{E}\|\sum_k Z_{t_k} [g(t_k, X^1_{t_k}) - g(t_k, X^2_{t_k})]\,\xi_k\|^2$$

$$= \mathbb{E}\|\sum_k \left(\int_0^1 g'(t_k, X^1_{t_k} + r(X^2_{t_k} - X^1_{t_k}))dr\right) \cdot Y_{t_k} \xi_k\|^2$$

$$\leq \mathbb{E}\left(\int_0^1 \|\sum_k g'(t_k, X^1_{t_k} + r(X^2_{t_k} - X^1_{t_k})) \cdot Y_{t_k} \xi_k\|^2\, dr\right)$$

$$\leq C \sum_k \mathbb{E}(Z_{t_k} \|X^2_{t_k} - X^1_{t_k}\|)^2 (t_{k+1} - t_k) .$$

This proves that $g \in \mathrm{Lip}_{\xi}(\mathbb{R}_+ \times E, F)$. ⌋

Remark: The proposition shows especially that the family $(g'(t,x))_{t \geq 0, x \in E}$ is contained in $S^2(E,F)$ and uniformly bounded relative to the norm σ_2.

4. EXAMPLES OF 2-SMOOTH OPERATORS

The results of section 3 indicate that the space $S^2(E,F)$ of 2-smooth operators from a Banach space E into a second Banach space F is an interesting space of operators, at least for the problem of stochastic integration of Banach space valued functions. In this section we will give a series of sufficient criteria for operators to be 2-smooth.

A well known subspace of $S^2(E,F)$ is the space $\Pi_2(E,F)$ of all 2-absolutely summing operators from E into F. If $T\in\Pi_2(E,F)$, and if (X_k) is an arbitrary sequence of E-valued, square integrable random vectors, adapted to the natural filtration $(\mathcal{F}_n)$ defined by a given Bernoulli sequence (ε_n), then $T\in S^2(E,F)$ follows from the inequality

$$\mathbb{E}\|\textstyle\sum_k TX_k\varepsilon_{k+1}\|^2 \leq \pi_2(T)^2 \sup\limits_{x'\in E',\|x'\|\leq 1} (\sum_k \mathbb{E}|\langle X_k,x'\rangle|^2)$$

$$\leq \pi_2(T)^2 \textstyle\sum_k \mathbb{E}\|X_k\|^2$$

The following example is also an immediate consequence of the defining inequality (2) of 3.5. Let E_1, F_1 denote further Banach spaces, and suppose that R,S,T are operators with $R\in L(F,F_1)$, $T\in L(E_1,E)$ and $S\in S^2(E,F)$. Then $R\circ S\circ T\in S^2(E_1,F_1)$ and $\sigma_2(R\circ S\circ T) \leq \|R\|\sigma_2(S)\|T\|$. Since a Banach space is 2-smoothable if and only if the identity operator is 2-smooth, we get $L(E,F) = S^2(E,F)$ in the case that E or F are 2-smoothable.

The spaces ℓ_p with $1\leq p<2$ are simple examples of not 2-smoothable Banach spaces and the next result illustrates the difference between $L(\ell_p)$ and $S^2(\ell_p)(:= S^2(\ell_p,\ell_p))$. Let $L_d(\ell_p)$ denote the space of diagonal operators on ℓ_p (i.e. operators T of the form $Te_i = c_i e_i$, $i = 1,2,\ldots$, where (e_i) is the unit vector basis in ℓ_p). $L_d(\ell_p)$ can be naturally identified with ℓ_∞. Let $S_d^2(\ell_p)$ denote the space of all 2-smooth diagonal operators. Then of course $S_d^2(\ell_p) = \ell_\infty = L_d(\ell_p)$, if $p\geq 2$. For $1\leq p<2$ we get:

4.1 <u>Proposition</u>. $S_d^2(\ell_p) = \ell_{\frac{2p}{2-p}}$ $(1\leq p<2)$.

Proof: Let (X,Σ,μ) be a σ-finite measure space and put $E := L^p(X,\Sigma,\mu)$. For an Σ-measurable function g let T_g denote the multiplication operator $T_g f = gf$. If g is bounded, then $T_g \in L(E)$. Now suppose that in addition $g \in L^{\frac{2p}{2-p}}(X,\Sigma,\mu)$, and let (X_k) be an $(\mathcal{G}_k)$-adapted sequence of square-integrable, E-valued random vectors. Then we obtain

$$\mathbb{E}\|\sum_k T_g X_k \varepsilon_{k+1}\|_p^2 = \mathbb{E}\left(\int|\sum_k gX_k\varepsilon_{k+1}|^p d\mu\right)^{2/p}$$

$$\leq \left(\int|g|^{\frac{2p}{2-p}} d\mu\right)^{\frac{2-p}{p}} \mathbb{E}\left(\int|\sum_k X_k \varepsilon_{k+1}|^2 d\mu\right)$$

$$\leq C \sum_k \mathbb{E}\left(\int|X_k|^2 d\mu\right) \quad \text{for} \quad C := \left(\int|g|^{\frac{2p}{2-p}} d\mu\right)^{\frac{2-p}{p}}.$$

For $X = \mathbb{N}$ this inequality implies $\ell_{\frac{2p}{2-p}} \subset S_d^2(\ell_p)$ for $1\leq p<2$ because of $\|\cdot\|_2 \leq \|\cdot\|_p$.

Now let $T = (c_i) \in S_d^2(\ell_p)$ and let (x_k) be an arbitrary sequence in ℓ_2. Define $X_k := x_k e_k$ for all $k\geq 1$. Then $T \in S_d^2(\ell_p)$ implies the inequality

$$\sum_k |c_k|^p |x_k|^p \leq C\left(\sum_k |x_k|^2\right)^{p/2}$$
$$= C\left(\sum_k (|x_k|^p)^{2/p}\right)^{p/2}$$

for a certain constant $C\geq 0$. Since this inequality is valid for every sequence (x_k) in ℓ_2, it follows that $(|c_i|^p) \in \ell_{q'}$, where q' is the conjugate exponent of $2/p$, i.e. $q' = \frac{2}{2-p}$. This shows $T \in \ell_{\frac{2p}{2-p}}$. $\lrcorner$

In the following we often make use of the following inequality (I_p) $(1\leq p<\infty)$, which for a given $T \in S^2(E,F)$ is equivalent to inequality (2) of corollary 3.5 (cf.[6], Remark 3.3).

$$(I_p) \quad \mathbb{E}\|\sum_k TX_k \varepsilon_{k+1}\|^p \leq C\,\mathbb{E}\left(\sum_k \|X_k\|^2\right)^{p/2}$$

for every $(\mathcal{G}_k)$-adapted sequence (X_k) of

square integrable E-valued random vectors. It is not difficult to see that there exists a general constant $D_p \geq 0$ such that (I_p) holds with the constant $C = D_p \sigma_2(T)^p$.

Let (E_k) be a sequence of Banach spaces with norms $\|\cdot\|_k$. Then the p-<u>direct</u> <u>sum</u> of (E_k) is defined as the space $(\sum^{\oplus} E_k)_p$ of all sequences (x_k) with $x_k \in E_k$ such that $(\|x_k\|_k) \in \ell_p$ $(1 \leq p \leq \infty)$.

4.2 <u>Proposition</u>. Let (E_k) and (F_k) be two sequences of Banach spaces, and let (T_k) be a sequence of operators $T_k \in S^2(E_k, F_k)$. Define the mapping $T: \prod_k E_k \longrightarrow \prod_k F_k$ by $T(x_k) := (T_k x_k)$ for every $(x_k) \in \prod_k E_k$. Then the following assertions hold:

(1) Let $1 \leq p \leq 2$. Then $T \in S^2((\sum^{\oplus} E_k)_2, (\sum^{\oplus} F_k)_p)$, if $(\sigma_2(T_k)) \in \ell_{\frac{2p}{2-p}}$.

(2) Let $2 \leq p \leq \infty$. Then $T \in S^2((\sum^{\oplus} E_k)_p, (\sum^{\oplus} F_k)_2)$ if $(\sigma_2(T_k)) \in \ell_{\frac{2p}{p-2}}$.

<u>Proof</u>: If $1 \leq p < 2$ and $(\sigma_2(T_k)) \in \ell_{\frac{2p}{2-p}}$, we use inequality (I_p). Let $(X_k) = ((X_{ki})_{i \geq 1})_{k \geq 1}$ be an arbitrary $(\mathcal{G}_k)$-adapted sequence of square integrable random vectors with values in $E := (\sum^{\oplus} E_k)_2$. Then we get

$$\mathbb{E}\|\sum_k TX_k \varepsilon_{k+1}\|^p = \mathbb{E}\left[\sum_i \|\sum_k T_i X_{ki} \varepsilon_{k+1}\|^p\right]$$

$$\leq \mathbb{E}\left[\sum_i D_p \sigma_2(T_i)^p (\sum_k \|X_{ki}\|^2)^{p/2}\right]$$

$$\leq D_p (\sum_i \sigma_2(T_i)^{\frac{2p}{2-p}})^{\frac{2-p}{2}} \mathbb{E}\left[\sum_k (\sum_i \|X_{ki}\|^2)\right]^{p/2}$$

since $\ell_{\frac{2}{2-p}}$ is the dual of $\ell_{\frac{2}{p}}$. The case p=2 is obtained similar.

Now let $2 < p < \infty$, and take a sequence (X_k) of square integrable random vectors in $E = (\sum^{\oplus} E_k)_p$ which is $(\mathcal{G}_k)$-adapted. Then we get

$$\mathbb{E}\|\sum_k TX_k \varepsilon_{k+1}\|^2 = \mathbb{E}\left[\sum_i \|\sum_k T_i X_{ki} \varepsilon_{k+1}\|^2\right]$$

$$\leq \mathbb{E}\left[\sum_i \sigma_2(T_i)^2 \sum_k \|X_{ki}\|^2\right]$$

$$\leq \left(\sum_i \sigma_2(T_i)^{\frac{2p}{p-2}}\right)^{\frac{p-2}{p}} \mathbb{E}\left[\sum_k \left(\left(\sum_i \|X_{ki}\|^p\right)^{1/p}\right)^2\right].$$

The cases $p=2, \infty$ are proved similar. ⌋

4.3 Corollary. Let (X,Σ,μ) be a σ-finite measure space. Then the following two assertions hold:

(1) If $1\leq p\leq 2$ and $T \in L^{2p/2-p}(X,\Sigma,\mu;S^2(E,F))$, then $T \in S^2(L^2(X,\Sigma,\mu;E),L^p(X,\Sigma,\mu;F))$.

(2) If $2\leq p\leq \infty$ and $T \in L^{2p/p-2}(X,\Sigma,\mu;S^2(E,F))$, then $T \in S^2(L^p(X,\Sigma,\mu;E),L^2(X,\Sigma,\mu;F))$.

The next result will be used in the last paragraph.

4.4 Proposition. Let (E_k) be a sequence of Banach spaces and let (T_k) be a sequence of operators $T_k \in S^2(E_k,F)$. If $\sum_k \|T_k\|^2 < \infty$, then $T(x_k) := \sum_k T_k x_k$ defines an operator $T \in L((\sum^{\oplus} E_k)_2, F)$. Suppose that T has the stronger property $\sum_k \sigma_2(T_k)^2 < \infty$. Then $T \in S^2((\sum^{\oplus} E_k)_2, F)$.

Proof: Let $(X_k) = ((X_{ki})_i)_k$ be an arbitrary $(\mathcal{G}_k)$-adapted, square integrable, $(\sum^{\oplus} E_k)_2$ -valued sequence of random vectors. Then the assertion follows from the following chain of inequalities:

$$\mathbb{E}\|\sum_k TX_k \varepsilon_{k+1}\|^2 = \mathbb{E}\|\sum_i \sum_k T_i X_{ki} \varepsilon_{k+1}\|^2$$

$$= \left(\sum_i \sigma_2(T_i)^2\right)^2 \mathbb{E}\left\|\sum_i \frac{\sigma_2(T_i)^2}{\sum_j \sigma_2(T_j)^2} \frac{T_i}{\sigma_2(T_i)^2} \sum_k X_{ki} \varepsilon_{k+1}\right\|^2$$

$$\leq \left(\sum_i \sigma_2(T_i)^2\right) \sum_i (\sigma_2(T_i))^{-2} \mathbb{E}\|T_i \sum_k X_{ki} \varepsilon_{k+1}\|^2$$

$$\leq \left(\sum_i \sigma_2(T_i)^2\right) \sum_i \mathbb{E}\left[\sum_k \|X_{ki}\|^2\right].$$ ⌋

4.5 Corollary. Let (X,Σ,μ) be a σ-finite measure space and suppose that $T \in L^2(X,\Sigma,\mu;S^2(E,F))$. Then $T \in S^2(L^2(X,\Sigma,\mu;E),F)$.

5. THE SPACE OF SMOOTHLY INTEGRABLE FUNCTIONS

We know from theorem 2.3 that in general the space $L^2(\lambda\otimes P,E)$ is not contained in $\mathbb{L}^2_\xi$. Motivated by the results of section 3 we introduce in this section a subspace of $L^2(\lambda\otimes P,E)$ which is always contained in $\mathbb{L}^2_\xi$ and which is at the same time rich enough to contain a sufficiently large class of functions, which can occur as integrands of stochastic integrals.

5.1 <u>Definition</u>. A function $f\in L^2(\lambda\otimes P,E)$ is called <u>smoothly integrable</u>, if there exist a Banach space G, a function $g\in L^2(\lambda\otimes P,G)$, and an operator $T\in S^2(G,E)$ such that $f=T\circ g$. The space of all smoothly integrable functions will be denoted by $L^2_s(\lambda\otimes P,E)$.

It follows immediately from the definition that $L^2_s(\lambda\otimes P,E)\subset\mathbb{L}^2_\xi(E)$ and that

$$\mathbb{E}\|\int f d\xi\|^2 \leq \sigma_2(T)^2\, \mathbb{E}\int\|g\|^2\, d\lambda \quad ,$$

if (G,T,g) is a triple occuring in definition 5.1 with $T\circ g = f$. Since this factorization of f is surely not unique, we introduce the following notation. For every $f\in L^2_s(\lambda\otimes P,E)$ we denote by F(f) the set of all triples (G,T,g) such that G is a Banach space, $T\in S^2(G,E)$ with $\sigma_2(T)\leq 1$, and $g\in L^2(\lambda\otimes P,G)$ such that $T\circ g = f$.

For every $f\in L^2_s(\lambda\otimes P,E)$ we define

$$\|f\|_s \;:=\; \inf_{(G,T,g)\in F(f)} \left(\mathbb{E}\int\|g\|^2 d\lambda\right)^{1/2} \quad .$$

Hence for all $f\in L^2_s(\lambda\otimes P,E)$ we have the inequality

$$\mathbb{E}\|\int f d\xi\|^2 \leq \|f\|_s^2 \quad .$$

5.2 <u>Lemma</u>. $\|\cdot\|_s$ is a norm on $L^2_s(\lambda\otimes P,E)$.

<u>Proof</u>: We prove that $\|\cdot\|_s$ is a homogeneous and convex function. Since the homogenity is obvious it remains to prove that $\|\cdot\|_s$ is convex on the set of all $f\in L^2_s(\lambda\otimes P,E)$ with $\|f\|_s\leq 1$.

So let $f_1,f_2\in L^2_s(\lambda\otimes P,E)$ with $\|f_1\|_s\leq 1$, $\|f_2\|_s\leq 1$ and $\lambda_1,\lambda_2\in\,]0,1[$ with $\lambda_1+\lambda_2 = 1$ be given. For every $\varepsilon > 0$ there exist triples $(G_1 T_1, g_1) \in F(f_1)$, $(G_2, T_2, g_2) \in F(f_2)$ such that

$$\mathbb{E}\int \| g_1\|^2 d\lambda \leq \| f_1\|_s^2 + \varepsilon \quad , \quad \mathbb{E}\int \| g_2\|^2 d\lambda \leq \| f_2\|_s^2 + \varepsilon$$

Consider the triple (G,T,g) with $G := (G_1 \oplus G_2)_2$ (the 2-direct sum, cf. section 4), $T := (\sqrt{\lambda_1} T_1, \sqrt{\lambda_2} T_2)$ and $g := (\sqrt{\lambda_1} g_1, \sqrt{\lambda_2} g_2)$. Then we have

$$T\,g = \lambda_1 f_1 + \lambda_2 f_2 \quad \text{and by proposition 4.4}$$

$$\sigma_2(T)^2 \leq \sigma_2(\sqrt{\lambda_1} T_1)^2 + \sigma_2(\sqrt{\lambda_2} T_2)^2 \leq 1 \; .$$

This implies

$$\| \lambda_1 f_1 + \lambda_2 f_2\|_s^2 \leq \mathbb{E}\int \| g\|^2 d\lambda = \mathbb{E}\int \| (\sqrt{\lambda_1} g_1, \sqrt{\lambda_2} g_2)\|^2 d\lambda$$

$$= \lambda_1 \, \mathbb{E}\int \| g_1\|^2 d\lambda + \lambda_2 \, \mathbb{E}\int \| g_2\|^2 d\lambda \leq 1 + \varepsilon,$$

and hence $\| \lambda_1 f_1 + \lambda_2 f_2\|_s \leq 1$ since $\varepsilon > 0$ was arbitrary.]

5.3 Theorem. $L_s^2(\lambda \otimes P, E)$ is a Banach space relative to the norm $\| . \|_s$.

Proof: We have to prove that every $\| . \|_s$-Cauchy sequence converges to an element in $L_s^2(\lambda \otimes P, E)$. Let $(f_n)_{n \geq 1}$ be a given $\| . \|_s$-Cauchy sequence. By definition of the norm $\| . \|_s$ this means that for every $\varepsilon > 0$ there is an n_ε such that for all $n, m \geq n_\varepsilon$ there exists a triple $(G_{n,m}, T_{n,m}, g_{n,m}) \in F(f_n - f_m)$ such that $\mathbb{E}\int \| g_{n,m}\|^2 d\lambda < \varepsilon$.

From this we get that there exists a strictly increasing sequence (n_k) in $\mathbb{N}$ and a sequence of triples $(G_k, T_k, g_k) \in F(f_{n_{k+1}} - f_{n_k})$ such that

$$\mathbb{E}\int \| g_k\|^2 d\lambda \leq 2^{-2(k+1)} \qquad \text{for all } k \geq 1.$$

Finally, we choose a triple $(G_0, T_0, g_0) \in F(f_{n_1})$

and put for every $k\geq 1$: $F_k := G_{k-1}$, $S_k := 2^{-k/2} T_{k-1}$ and $h_k := 2^{k/2} g_{k-1}$.

Then we define

$$F := (\sum^{\oplus}_{k\geq 1} F_k)_2 ,$$

$$S := (S_k)_{k\geq 1} \text{ and } h := (h_k)_{k\geq 1} .$$

It follows from proposition 4.4 that

$$\sigma_2(S)^2 \leq \sum_k \sigma_2(S_k)^2 = \sum_k 2^{-k}\sigma_2(T_{k-1})^2 \leq 1 .$$

This implies that $(F,S,h) \in F(f)$ for $f := S\circ h$.

By definition of S and f we get for every $k\geq 1$:

$$\|f - f_{n_k}\|_s^2 = \|f - \sum_{j=1}^{k-1}(f_{n_{j+1}} - f_{n_j}) - f_{n_1}\|_s^2$$

$$= \|S\circ h - \sum_{j=1}^{k} S_j h_j\|_s^2 = \|\sum_{j\geq k+1} S_j h_j\|_s^2$$

$$\leq \sum_{j\geq k+1} \mathbb{E}\int \|h_j\|^2 d\lambda = \sum_{j\geq k} 2^{j+1}\, \mathbb{E}\int \|g_j\|^2 d\lambda$$

$$\leq \sum_{j\geq k} 2^{-(j+1)} = 2^{-k}.$$

This proves that $\lim_{n\to\infty} \|f - f_n\|_s = 0$. ⌋

We close this paragraph in giving a general necessary and sufficient condition that a deterministic function belongs to L_s^2. This criterion may serve as an additional motivation to consider L_s^2 as a best space of functions for which the stochastic integral exists.

5.4 <u>Theorem</u>. For a function $f: \mathbb{R}_+ \longrightarrow E$ the following assertions are equivalent:

(1) $f \in L_s^2(\lambda\otimes P, E)$.

(2) There exists a function $g: \mathbb{R}_+ \longrightarrow \mathbb{R}_+$ such that for all $Z, U \in \mathcal{E}(\mathbb{R})$, $U \geq 0$ the following inequality holds:

$$\mathbb{E}\left\| \int Z\, f(U)\, d\xi \right\|^2 \leqslant \int \mathbb{E}\left[Z^2 g(U)\right] d\lambda \ .$$

Proof: If (1) holds, then there exist a Banach space G, a function $h : \mathbb{R}_+ \longrightarrow G$ and a $T \in S^2(G,E)$ such that $f = T \circ h$. Then the inequality in (2) holds with the function $g := \sigma_2(T)^2 \|h\|^2$.

Now suppose that (2) holds. We proceed similar as in the proof of lemma 3.2. For all $s \geqslant o$ we define for all $y \in E$:

$$n_s(y)^2 := \sup_{\substack{Z \in \mathcal{E}_s(\mathbb{R}) \\ U \in \mathcal{E}_s(\mathbb{R}),\, U \geqslant 0}} \left(\mathbb{E}\left\| y + \int_s^\infty Z f(U)\, d\xi \right\|^2 - 2\int \mathbb{E}\left[Z^2 g(U)\right] d\lambda \right)$$

Then for every $s \geqslant o$, n_s is an equivalent norm on E such that $n(y) := n_o(y) = \lim_{s \downarrow o} n_s(y)$ for all $y \in E$. Exactly as in the proof of lemma 3.2 one can first show that for every $h > o$, $t \geqslant o$ the inequality

$$\mathbb{E}\, n_h(y + f(t)\xi_h)^2 \leqslant n(y)^2 + hg(t)$$

holds for every $y \in E$. Dividing this inequality by h we obtain for $h \downarrow o$

$$(*) \quad \mathbb{E}\, n(y + f(t)\xi)^2 \leqslant n(y)^2 + g(t)$$

for all $y \in E$, $t \geqslant o$, where ξ is $N(o,1)$-distributed. As in the proof of theorem 3.7 we introduce the Banach space G of all $x \in E$ such that

$$m(x) := \sup_{y \in E} \left[\mathbb{E}\, n(y + x\xi)^2 - n(y)^2\right]^{1/2} < \infty \ .$$

Then the injection T from G into E is 2-smooth. By inequality (*) the function f is G-valued. This proves $f \in L_s^2(\lambda \otimes P, E)$. ⌋

Egbert Dettweiler
Mathematisches Institut
der Universität Tübingen
Auf der Morgenstelle 1o
74oo Tübingen
West Germany

REFERENCES

1. Dettweiler, E.: Banach space valued processes with independent increments and stochastic integration. In: Probability in Banach spaces IV, Proceedings, Oberwolfach 1982, Lecture Notes in Math. 99o, 54-83. Berlin-Heidelberg-New York: Springer 1983.
2. Dettweiler, E.: Stochastic Integral Equations and Diffusions on Banach Spaces. To appear in: Probability Theory on Vector Spaces, Proceedings of the third int. Conf., Lublin 1983.
3. Dynkin, E.B.: Markov Processes. Vol.I&II. Berlin-Göttingen-Heidelberg: Springer 1965.
4. Métivier, M.: Semimartingales. Berlin-New York: de Gruyter 1982.
5. Métivier, M., Pellaumail, J.: Stochastic Integration. New York: Academic Press 198o.
6. Pisier, G.: Martingales with values in uniformly convex spaces. Israel J. Math. 2o, 326-35o (1975).
7. Woyczynski, W.A.: Geometry and Martingales in Banach Spaces. In: Winter School on Probability, Karpacz 1975, 229-275. Lecture Notes in Math. 472. Berlin-Heidelberg-New York: Springer 1975.

Akira Ichikawa

A SEMIGROUP MODEL FOR PARABOLIC EQUATIONS WITH BOUNDARY AND POINTWISE NOISE

1. Introduction.

Recently the semigroup model for controlled partial differential equations and stochastic partial differential equations has been adopted by many researchers in system theory see [2] and its references. When control or noise has the so-called distributed nature, then abstract operators associated with them are bounded and the system theory can be developed almost as in finite dimensions. However, there are some new aspects in infinite dimensional systems. One example is a boundary input (control).

Let $\mathcal{O}$ be a bounded open domain in R^d with boundary Γ. We assume that Γ is a d-1 dimensional variety and that $\mathcal{O}$ lies on one side of Γ. Let $A(x,\partial) = \sum_{|m|\leq 2} a_m \partial^m$ be a uniformly strongly elliptic operator of order two with smooth coefficients $a_m(x)$. Consider a nonhomogeneous boundary value problem of mixed type:

$$\begin{aligned} &\partial y(x,t)/\partial t = -A(x,\partial)y(x,t), \\ &y(x,0) = y_0(x), \\ &\partial y(x,t)/\partial n + a(x)y(x,t) = g(x)u(t) \quad \text{on } \Gamma, \end{aligned} \tag{1.1}$$

where $y_0(x) \in L_2(\mathcal{O})$, $g \in L_2(\Gamma)$, ∂ / ∂_n is the outward normal derivative, $a(x)$ is a real positive function defined on Γ and $u(t)$ is a locally integrable function. Define $D(-A)$ as the closure in $H^2(\mathcal{O})$ of the subspace of $C^2(\overline{\mathcal{O}})$ which consists of functions ϕ satisfying the boundary condition $\partial\phi/\partial n + a\phi = 0$. Let $-A$ be the restriction of $-A(x,\partial)$ on $D(-A)$. Then $-A$ generates an analytic semigroup $S(t)$ [17]. Let M be the map: $L_2(\Gamma) \to L_2(\mathcal{O})$ defined by $y = Mg$, where y is the solution of $-A(x,\partial)y = 0$, $\partial y/\partial n + ay = g$. It is known [11], [12], [15], [16] that (1.1) can be described by the semigroup model

$$y(t) = S(t)y_0 + \int_0^t AS(t-r)Mgu(r)dr. \tag{1.2}$$

L. Arnold and P. Kotelenez (eds.), Stochastic Space-Time Models and Limit Theorems, 81–94.

Now we take a feedback input i.e., a function of y,

$$u = f(y) \tag{1.3}$$

where f is a real continuous function on some subspace of $L_2(\mathcal{O})$. Then (1.2) becomes an integral equation

$$y(t) = S(t)y_0 + \int_0^t AS(t-r)Mgf(y(r))dr. \tag{1.4}$$

Then new feature of this model is that the generator A appears in the integral term. The existence and uniqueness problem for systems of type (1.4) has been studied by Lasiecka and Triggiani [12].

Now we replace (1.3) by

$$u = f(y) + kf(y)\dot{w}(t) \tag{1.5}$$

where $k > 0$ and $\dot{w}(t)$ is a white noise. Then (1.4) is formally written

$$y(t) = S(t)y_0 + \int_0^t AS(t-r)Mgf(y(r))dr + k\int_0^t AS(t-r)Mgf(y(r))\dot{w}(r)dr$$

If we replace the last integral by the Ito integral, we obtain

$$y(t) = S(t)y_0 + \int_0^t AS(t-r)Mgf(y(r))dr + k\int_0^t AS(t-r)Mgf(y(r))dw(r), \tag{1.6}$$

where w(t) is the standard Wiener process. We may regard this as an abstract model for a parabolic equation with boundary noise. The stochastic evolution equation of this type is new.

We assume that the fractional powers A^γ, $0 < \gamma < 1$, are well defined. Then $Mg \in \mathcal{D}(A^{1-\gamma})$ for any $\gamma > 1/4$ [5], [12], [14]. Now set $z(t) = A^{-\gamma}y(t)$, then (1.6) yields

$$z(t) = S(t)z_0 + \int_0^t S(t-r)bh(z(r))dr + k\int_0^t S(t-r)bh(z(r))dw(r), \tag{1.7}$$

where $z_0 = A^{-\gamma}y_0$, $b = A^{1-\gamma}Mg$ and $h(z) = f(A^\gamma z)$. This is equivalent to (1.6), but it can describe a wider class of systems. So we shall first establish existence, uniqueness and regularity results for (1.7) and then translates them to (1.6). It turns out that we can apply our abstract results not only to systems with boundary noise but also to systems with point noise. Applications will be discussed in section 3.

Stochastic evolution equations of type (1.7) have been considered by Ichikawa [8], Da Prato [3] and Da Prato, Iannelli and Tubaro [4] but not from the point of view of boundary noise. Curtain [1] has proposed a semigroup model for systems with boundary noise and it corresponds to the special case where f(y) is independent of y. Zabczyk [16] gives a model in which boundary values satisfy stochastic differential equations. His model falls in the standard class [2], although the state space is larger.

2. Existence and uniqueness.

Let S(t) be a strongly continuous analytic semigroup on a real separable Hilbert space Y and -A its infinitesimal generator. We assume that fractional powers A^γ, $0 < \gamma < 1$ are well defined and that for each $0 < T < \infty$ there exists a constant $K = K(T) > 0$ such that

$$|A^\gamma S(t)| \leq K/t^\gamma \quad \text{for any} \quad 0 < \gamma < 1 \quad \text{and} \quad t \in (0,T]. \tag{2.1}$$

Here $|\,|$ denotes the norm of bounded linear operators on Y.

We consider the stochastic integral equation

$$z(t) = S(t)z_0 + \sum_{i=1}^{J}\int_0^t S(t-r)b_{0i}h_{0i}(z(r))dr$$
$$+ \sum_{i=1}^{N}\int_0^t S(t-r)b_i h_i(z(r))dw_i(r), \tag{2.2}$$

where z_0, b_{0i}, $b_i \in Y$, h_{0i} and h_i are Lipschitz continuous functions on $\mathcal{D}(A^\alpha)$ for some $0 < \alpha < 1/2$, J and N are positive integers and $w(t) = (w_i(t))$ is an N-dimensional Wiener process defined on a probability space $(\Omega,\mathcal{F},P)$. We denote by $\mathcal{F}_t$, the σ-algebra generated by w(s), $0 \leq s \leq t$.

We have the following existence result.

Theorem 2.1. For each $\alpha \leq \beta < 1/2$ there exists a unique $\mathcal{F}_t$-adapted solution $z(\cdot)$ to (2.2) in $L_p((0,T)\times\Omega;\mathcal{D}(A^\beta))$ for any $2 \leq p < 1/\beta \leq 1/\alpha$. For each $0 \leq \bar{\beta} < \alpha$, the solution $z(\cdot)$ $\varepsilon\ L_p((0,T)\times\Omega;D(A^{\bar{\beta}}))$ for any $2 \leq p < 1/\alpha$. Moreover, if $h_{0i}(0) = h_i(0) = 0$, then

$$E|z(t)|_\beta^p \leq M|z_0|^p/t^{\beta p} \quad \text{a.e.t for some } M = M(p,T) > 0 \quad (2.3)$$

where $||_\beta$ is the graph norm on $D(A^\beta)$.

Proof. For simplicity we shall take $b_{0i} = 0$, $N = 1$ and assume that $h = h_1$ is linear. The general case is an easy modification of this case. Then (2.2) becomes

$$z(t) = S(t)z_0 + \int_0^t S(t-r)bh(z(r))dw(r). \quad (2.4)$$

Let V be the closed subspace of $L_p((0,T)\times\Omega;D(A^\beta))$ which consists of F_t adapted processes. For each $z(\cdot)\ \varepsilon\ V$ define $(\Lambda z)(t)$ by the right hand side of (2.4). We shall show that Λ maps V into itself. In fact from (2.1) we have

$$|S(t)z_0|_\beta^p = |A^\beta S(t)z_0|^p \leq K^p|z_0|^p/t^{p\beta}\ \varepsilon\ L_1(0,T),$$

$$E|\int_0^t S(t-r)bh(z(r))dw(r)|_\beta^p \leq C(c_0K|b|)^p\int_0^t [1/(t-r)^{p\beta}]E|z(r)|_\beta^p dr \quad \varepsilon\ L_1(0,T),$$

where c_0 is the norm of f and $C = C(p,T) > 0$ is a constant. Choosing a suitable equivalent norm in $L_p((0,T)\times\Omega;D(A^\beta))$ as in [7], we can show that Λ is a contraction on V. Thus we have a unique solution in V. From (2.4) we have

$$E|z(t)|_\beta^p \leq (2K|z_0|)^p/t^{p\beta} + C(2c_0K|b|)^p\int_0^t \{E|z(r)|_\beta^p/(t-r)^{p\beta}\}dr.$$

Thus by the Gronwall's inequality below [6, p.6] we obtain (2.3). The second part follows similarly from the estimates of type above.

Lemma 2.1. Let $0 \leq \alpha, \beta < 1$, $0 < T < \infty$, $a \geq 0$, $b \geq 0$. If $u: [0,T] \to R^1$ is integrable such that

$$0 \leq u(t) \leq at^{-\alpha} + b\int_0^t (t-s)^{-\beta} u(s)ds \quad \text{a.e. } t \in [0,T],$$

then

$$0 \leq u(t) \leq aMt^{-\alpha} \quad \text{a.e. on } [0,T],$$

for some $M = M(b,\alpha,\beta,T) < \infty$.

Now we consider regularity of solutions.

Theorem 2.2. For each $0 \leq \delta < 1/2$, $z(\cdot) \in C((0,T];L_q(\Omega;D(A^\delta)))$ for any q, $2 \leq q < 1/\delta$ if $\alpha \leq \delta$ and for any $2 \leq q < 1/\alpha$ if $0 < \delta < \alpha$.
If, in particular, $z_0 \in D(A^\delta)$, $0 \leq \delta < 1/2$, then $z(\cdot) \in C([0,T];L_q(\Omega;D(A^\delta)))$ for any q with $2 \leq q < 1/\delta$ if $\alpha \leq \delta$ and $2 \leq q < 1/\alpha$ if $\delta < \alpha$.

Proof. First we shall show that $E|z(t)|_\delta^q$ is well defined for each $t > 0$. We shall only consider the case $\alpha \leq \delta < 1/2$. In fact we have

$$\begin{aligned} E\left|\int_0^t S(t-r)bh(z(r))dw(r)\right|_\delta^q &\leq C_1\int_0^t \{E|z(r)|_\delta^q/(t-r)^{q\delta}\}dr \\ &\leq C_1 M|z_0|^q \int_0^t dr/r^{q\delta}(t-r)^{q\delta} \qquad (2.5) \\ &< \infty \quad \text{for all } t \in (0,T], \end{aligned}$$

where C_1 and M are some positive constants. Hence $E|z(t)|_\delta^q$ is finite for all $t \in (0,T]$. Now we shall show the continuity of $z(t)$ in $L_q(\Omega;D(A^\delta))$ for each $t \in (0,T]$. Let $t > s > 0$ and consider

$$\begin{aligned} z(t) - z(s) = S(t)z_0 - S(s)z_0 &+ \int_s^t S(t-r)bh(z(r))dw(r) \\ &+ \int_0^s [S(t-r) - I]S(s-r)bh(z(r))dw(r). \end{aligned}$$

First we show

$$|S(t)z_0 - S(s)z_0|_\delta \to 0 \text{ as } t \downarrow s > 0 \text{ or as } s \uparrow t > 0.$$

Note that

$$S(t)z_0 - S(s)z_0 = \int_s^t AS(r)z_0 dr.$$

Thus

$$\begin{aligned}|S(t)z_0 - S(s)z_0|_\delta &= |A^\delta \int_s^t AS(r)z_0 dr| \\ &\leq \int_s^t |A^{1+\delta}S(r)z_0| dr \\ &\leq \int_s^t (\overline{K}/r^{1+\delta})|z_0| dr, \ \overline{K} > 0 \\ &= (\overline{K}/\delta)|z_0|(1/s^\delta - 1/t^\delta) \\ &\to 0 \text{ as } t \downarrow s > 0 \text{ or as } s \uparrow t > 0.\end{aligned}$$

Consider now

$$\begin{aligned}E|\int_s^t S(t-r)bh(z(r))dw(r)|_\delta^q &\leq C_2|z_0|^q \int_s^t [1/r^{q\delta}(t-r)^{q\delta}]dr, \ C_2 > 0, \\ &\leq (C_2|z_0|^q/(1-q^\delta)s^{q\delta})(t-s)^{1-q\delta} \\ &\to 0 \text{ as } t \downarrow s > 0 \text{ or } s \uparrow t > 0.\end{aligned}$$

Finally we consider

$$\begin{aligned}&E|\int_0^s [S(t-s) - I]S(s-r)bh(z(r))dw(r)|_\delta^q \\ &\quad \leq C\int_0^s |[S(t-s) - I]S(s-r)b|_\delta^q E|z(r)|_\delta^q dr\end{aligned}$$

$$\leq C\int_0^s \left|\int_0^{t-s} A^{1+\delta}S(u+s-r)bdu\right|_\delta^q E|z(r)|_\delta^q dr$$

$$\leq C(|b|/\delta)^q M\int_0^s |1/(s-r)^\delta - 1/(t-r)^\delta|/r^{q\delta}dr$$

$$\to 0 \text{ as } t \downarrow s > 0 \text{ or as } s \uparrow t > 0.$$

Hence we have shown that

$$E|z(t) - z(s)|_\delta^q \to 0 \text{ as } t \downarrow s > 0 \text{ or as } s \uparrow t > 0.$$

Now let $z_0 \in D(A^\delta)$. Then clearly $S(t)z_0 \in C([0,T];D(A^\delta))$. We can easily see

$$E\left|\int_0^t S(t-r)bh(z(r))dw(r)\right|_\delta^q \to 0 \text{ as } t \to 0$$

from the estimate of the form (2.5). Hence $z(\cdot) \in C([0,T]; L_q(\Omega;D(A^\delta)))$.

Remark 2.1. $z(\cdot) \in L_p((0,T)\times\Omega; (A^\beta))$ for any $\beta < 1/2$ and $1 \leq p < 2$. The conclusion of the Theorem 2.2 is also true for any $1 \leq q < 2$.

Remark 2.2. If $z_0 \in D(A^\beta)$ for some $\alpha \leq \beta < 1/2$, then we can develop existence in $C([0,T];L_q(\Omega;D(A^\beta)))$ similar to [4].

Remark 2.3. Existence results in L_p space for the deterministic case is given in [10].

Lastly we consider the continuity of sample paths.

Theorem 2.3. Let $0 \leq \delta < 1/2$. Suppose z_0, $b \in D(A^\delta)$. Then $z(t)$ has a version with continuous sample paths in $D(A^\delta)$.

Proof. Define $z(t) = A^\delta z(t)$, then

$$\overline{z}(t) = S(t)A^{\delta}z_0 + \int_0^t S(t-r)A^{\delta}bh(z(r))dw(r).$$

Thus $z(t)$ is continuous a.s. in $\mathcal{D}(A^{\delta})$ if and only if $\overline{z}(t)$ is continuous in Y a.s. But since $A^{\delta}z_0$, $A^{\delta}b \in Y$ and $-A$ is dissipative, by Proposition 3.8 [9] $\overline{z}(t)$ has a version with continuous sample paths in Y.

3. Applications.

3.1. Noise depending on the value at a point. Consider the stochastic parabolic equation

$$\begin{aligned} &dz(x,t) = -A(x,\partial)z(x,t)dt + b(x)z(x_0,t)dw(t), \ x,x_0 \in \mathcal{O} \\ &z(x,0) = z_0(x), \ x \in \mathcal{O}, \qquad\qquad (3.1) \\ &z(x,t) = 0 \quad \text{on} \quad \Gamma, \end{aligned}$$

where $\mathcal{O}$, Γ and $A(x,\partial)$ are given as in section 1. In this case $\mathcal{D}(-A) = H^2(\mathcal{O}) \cap H_0^1(\mathcal{O})$. Since $H^s(\mathcal{O}) \subset C(\overline{\mathcal{O}})$ for $s > d/2$ and $\mathcal{D}(A^s) \subset H^{2s}(\mathcal{O})$, we have $\mathcal{D}(A^{\alpha}) \subset C(\overline{\mathcal{O}})$ if $\alpha > d/4$. Hence, if $d \geq 2$, the condition of Theorem 2.1 is not satisfied. So we assumed $d = 1$. Then we can take $\alpha = 1/4 + \varepsilon$ for any $\varepsilon > 0$. And for any $1/4 < \beta < 1/2$ we have a unique F_t-adapted solution $z \in L_p((0,T)\times\Omega;\mathcal{D}(A^{\beta}))$ for $2 \leq p < 1/\beta$ to

$$z(t) = S(t)z_0 + \int_0^t S(t-r)bh(z(r))dw(r),$$

where $h(z) = z(x_0)$ is a continuous linear functional on $\mathcal{D}(A^{\beta})$, $\beta > 1/4$. We can take $\beta = 1/2 - \varepsilon$ for any small $\varepsilon > 0$. Then $z \in L_{2/(1-\varepsilon)}((0,T)\times\Omega;\mathcal{D}(A^{1/2-\varepsilon}))$. More suggestively we write this as $z(\cdot) \in L_{2+}((0,T)\times\Omega;\mathcal{D}(A^{1/2-}))$. Similarly we have $z(\cdot) \in L_{4-}((0,T)\times\Omega;\mathcal{D}(A^{1/4+}))$. From Theorem 2.2 we have for any $0 \leq \delta < 1/2$, $z(\cdot) \in C((0,T];L_q(\Omega;(A^{\delta}))$ for any q with $2 \leq q < 1/\delta$ if $\delta \geq 1/4 + \varepsilon$, and $2 \leq q < 4$ if $\delta < 1/4 + \varepsilon$. Hence

$$z(\cdot) \in C((0,T];L_{2+}(\Omega;D(A^{1/2-}))) \cap C((0,T];L_{4-}(\Omega;D(A^{1/4})).$$

If, in particular, $z_0 \in D(A^\delta)$, $0 \le \delta < 1/2$, then $z(\cdot) \in C([0,T];L_q(\Omega,D(A^\delta)))$ for any q with $2 \le q < 1/\delta$ if $\alpha \le \delta$ and $2 \le q < 1/\alpha$ if $\delta < \alpha$.
Since $z_0 \in D(A^0) = Y$ we have $z(\cdot) \in C([0,T];L_{4-}(\Omega;Y))$.
Summing up we have

$$z(\cdot) \in C((0,T];L_{2+}(\Omega;D(A^{1/2-}))) \cap C((0,T];L_{4-}(\Omega;D(A^{1/4+})))$$
$$\cap\, C([0,T];L_{4-}(\Omega;Y))).$$

If z_0, $b \in D(A^\delta)$, $0 \le \delta < 1/2$, then from Theorem 2.3 z(t) has a version with continuous sample paths in $D(A^\delta)$. In particular z(t) has alsways a version with continuous sample paths in $Y = L_2(O)$.

3.2. Boundary noise: mixed boundary condition. Consider

$$\begin{aligned} &\partial y(x,t)/\partial t = -A(x,\partial)y(x,t),\ x \in O,\ t > 0, \\ &y(x,0) = y_0(x),\ x \in O, \\ &\partial y(x,t)/\partial n + a(x)y(x,t) = g(x)f(y(\cdot,t))\dot{w}(t),\ x \in \Gamma, \end{aligned} \tag{3.2}$$

where $\dot{w}(t)$ is a white noise and f is a real function specified later. As we have seen in section 1, the semigroup model for (3.2) is

$$y(t) = S(t)y_0 + \int_0^t A^\nu S(t-r)A^{1-\nu}Mgf(y(r))dw(r) \tag{3.3}$$

for $\nu = 1/4 + \varepsilon$, since $Mg \in D(A^{3/4-\varepsilon})$ for any $\varepsilon > 0$. The z model corresponding to (3.3) is

$$z(t) = S(t)z_0 + \int_0^t S(t-r)bf(A^\nu z(r))dw(r),\ z_0 = A^{-\nu}y_0 \in D(A^\nu).$$

In order to satisfy the condition of Theorem 2.1, we can take

continuous functions f on $\mathcal{D}(A^{\delta})$ only for $0 \leq \delta < 1/4$.
Case(i): $f(y) = c + \langle f,y\rangle$, where c is a constant and $f \in Y$. Then we can take $\alpha = \nu = 1/4 + \varepsilon$. Hence by Theorems 2.1 and 2.2 we have

$$z(\cdot) \in L_{2+}((0,T)\times\Omega;\mathcal{D}(A^{1/2-}))) \cap L_{4-}((0,T)\times\Omega;\mathcal{D}(A^{1/4+}))$$
$$\cap\, C((0,T];L_{2+}(\Omega;A^{1/2-})))$$
$$\cap\, C([0,T];L_{4-}(\Omega;(A^{1/4+}))).$$

Thus translating this to y(t) we have

$$y(\cdot) \in L_{2+}((0,T)\times\Omega;\mathcal{D}(A^{1/4-})) \cap C((0,T];L_{2+}(\Omega;\mathcal{D}(A^{1/4-})))$$
$$\cap\, C([0,T];L_{4-}(\Omega;Y)).$$

Since we cannot expect b in any $\mathcal{D}(A^{\delta})$, $\delta > 0$, we have sample continuity of z only in Y. Hence $A^{-\alpha}y(t)$ has continuous sample paths. Let Y_δ be the completion of Y with respect to the inner product $\langle\langle y,z\rangle\rangle = \langle A^{-\delta}y,A^{-\delta}z\rangle$, $y,z \in Y$. Then $y(\cdot)$ has continuous sample paths in $Y_{1/4+}$ where $Y_{\delta+}$ means $Y_{\delta+\varepsilon}$ for any $\varepsilon > 0$ small.
Case (ii): $f(y) = \langle f,y\rangle_\Gamma$, $f \in L_2(\Gamma)$. In this case f(y) is continuous on $\mathcal{D}(A^{\delta})$ for $\delta > 1/4$. Thus $h(z) = f(A^{\nu}z)$ is continuous on $\mathcal{D}(A^{\alpha})$ only for $\alpha > 1/2$. So the assumption of Theorem 2.1 is not satisfied.

3.3. Boundary noise: Dirichlet boundary condition. Consider

$$\begin{aligned} &\partial y(x,t)/\partial t = -A(x,\partial)y(x,t),\ x \in \mathcal{O},\ t > 0, \\ &y(x,0) = y_0(x) \\ &y(x,t) = g(x)f(y(\cdot,t))\dot{w}(t),\ x \in \Gamma, \end{aligned} \tag{3.4}$$

where $\mathcal{O}$, Γ and $A(x,\partial)$ are given as in section 1, $g \in L_2(\Gamma)$ and f and w(t) are as in (3.3).

Let D be the Dirichlet operator: $L_2(\Gamma) \to L_2(\mathcal{O})$ defined by

$y = Dg$: $-A(x,\partial)y(x) = 0$, $y(x)|_\Gamma = 0$.

Then $Dg \in \mathcal{D}(A^{1-\nu})$ for $\nu = 3/4 + \varepsilon$, for small $\varepsilon > 0$. Thus the semigroup model for (3.4) is

$$y(t) = S(t)y_0 + \int_0^t A^{\nu}S(t-r)A^{1-\nu}Dgf(y(r))dw(r). \qquad (3.5)$$

The z model associated with (3.5) is

$$z(t) = S(t)z_0 + \int_0^t S(t-r)bh(z(r))dw(r), \qquad (3.6)$$

where $b = A^{1-\nu}Dg$, $z_0 = A^{-\nu}y_0 \in \mathcal{D}(A^{\nu})$ and $h(z) = f(A^{\nu}z)$.
Hence the condition of Theorem 2.1 is not in general satisfied. However, if we take $f(y) = \langle f,y\rangle$ with $f \in \mathcal{D}(A^{*\delta})$ and $\delta > 1/4$, then $h(z)$ is continuous on $\mathcal{D}(A^{\alpha})$, $\alpha = \nu - \delta < 1/2$. Thus we can apply Theorems 1 and 2 and obtain solutions for (3.4). We take $\delta = \nu$ so that $h(z)$ is continuous on Y. For example $f \in \mathcal{D}(A^*)$. Then

$$z(\cdot) \in L_{2+}((0,T)\times\Omega;\mathcal{D}(A^{1/2-})) \cap C((0,T];L_{4-}(\Omega;\mathcal{D}(A^{1/4+})))$$
$$\cap\, C([0,T];L_q(\Omega;Y)) \text{ for any } 2 \le q < \infty.$$

Thus $y(\cdot) = A^{\nu}z(\cdot) \in L_{2+}((0,T)\times\Omega;Y_{1/4+}) \cap C([0,T];L_q(\Omega;Y_{3/4+}))$.
Moreover, y has continuous sample paths in $Y_{3/4+}$.

3.4. Pointwise noise. Consider

$$\begin{aligned} &dy(x,t) = -A(x,\partial)y(x,t)dt + \delta(x-x_0)f(y(\cdot,t))dw(t),\ x_0 \in \mathcal{O},\\ &y(x,0) = y_0(x),\ x \in \mathcal{O}, \qquad (3.7)\\ &\partial y(x,t)/\partial n + a(x)y(x,t) = 0,\ x \in \Gamma, \end{aligned}$$

where $\mathcal{O}$, Γ and $A(x,\partial)$ are given as in section 1. Let G be the Green's function defined by

$$-A(x,\partial)G(x,\xi) = \delta(x-\xi),\ \partial G(x,\xi)/\partial n + a(x)G(x,\xi) = 0,\ x \in \Gamma.$$

Let $g(x) = G(x,x_0)$. Then the semigroup model for (3.7) is the following:

$$y(t) = S(t)y_0 + \int_0^t AS(t-r)gf(y(r))dw(r). \tag{3.8}$$

We assume that A is self-adjoint, then $A^{1-\gamma}g \in Y$ for $\gamma > d/4$. So we take $d = 1$. Thus the z model for (3.8) is

$$z(t) = S(t)z_0 + \int_0^t S(t-r)bh(z(r))dw(r),$$

where $z_0 = A^{-\alpha}y_0$, $\alpha = 1/4 + \varepsilon$, $b = A^{1-\alpha}g$ and $h(z) = f(A^{\alpha}z)$. Hence for any Lipschitz continuous function f on $\mathcal{D}(A^{\gamma})$, $0 \leq \gamma < 1/4$, We apply Theorems 1 and 2. Let $f(y) = \langle f,y\rangle$, $f \in Y$, then we have

$$z(\cdot) \in L_{2+}((0,T)\times\Omega;\mathcal{D}(A^{1/2-})) \cap C([0,T];L_{4-}(\Omega;\mathcal{D}(A^{1/4+})))$$

and hence

$$y(\cdot) \in L_{2+}((0,T)\times\Omega;\mathcal{D}(A^{1/4-})) \cap C([0,T];L_{4-}(\Omega;Y)).$$

Moreover, $y(\cdot)$ has continuous sample paths in $Y_{1/4+}$. A possible choice of A is $-A = d^2/dx^2 - 1$, $\mathcal{D}(A) = \{y \in H^2(0,1),\ y'(0) = y'(1) = 0\}$. If $d = 2$, we can assume that $h(z)$ is continuous on $\mathcal{D}(A^{\gamma})$ for some $\gamma < 1/2$ and obtain solutions as in subsection 3.3. We may change the mixed boundary condition to the Dirichlet condition. Then for $d \leq 2$, we obtain solutions in a similar manner.

Faculty of Engineering,
Shizuoka University,
Hamamatsu 432, Japan

References

1. Curtain, R.F.: 1979, 'Stochastic distributed systems with point observation and boundary control: An abstract theory, Stochastics', 3, 85-104.
2. Curtain, R.F. and A. J. Pritchard: 1978, Infinite Dimensional Linear Systems Theory, Lecture Notes in Control and Information Sciences 8, Springer-Verlag, Berlin.
3. Da Prato, G.: 1983, 'Some results on linear stochastic evolution equations in Hilbert spaces by semigroup method', Stoch. Anal. Appl., 1, 57-88.
4. Da Prato, G. and M. Iannelli and L. Tubaro: 1979, 'Semilinear stochastic differential equations in Hilbert spaces', Bollettino U.M.I. (5) 16-A, 168-185.
5. Fujiwara, D.: 1967, 'Concrete characterization of the domains of fractional powers of some elliptic differential operators of the second order', Proc. Japan Acad. 43, 82-86.
6. Henry, D.: 1981, Geometric Theory of Semilinear Parabolic Equations, Lecture Notes in Math., 840, Springer-Verlag, Berlin.
7. Ichikawa, A.: 1978, 'Linear stochastic evolution equations in Hilbert space', J. Diff. Eqns., 28, 266-277.
8. Ichikawa, A.: 1980, 'Stability and control of stochastic evolution equations', Analysis and Optimization of Stochastic Systems, O.L.R. Jacobs, et. al. (ed.), Academic Press, London.
9. Ichikawa, A.: 1984, 'Semilinear stochastic evolution equations: Boundedness, stability and invariant measures', Stochastics, 12, 1-39.
10. Ichikawa, A. and A. J. Pritchard: 1979, 'Existence, uniqueness and stability of nonlinear evolution equations', J. Math. Anal. Appl., 68, 454-476.
11. Lasiecka, I.: 1980, 'Unified theory for abstract parabolic boundary problems - A semigroup approach', Appl. Math. Optim., 6, 281-333.
12. Lasiecka, I. and R. Triggiani: 1983, 'Feedback semigroups and cosine operators for boundary feedback parabolic and hyperbolic equations', J. Diff. Eqns., 47, 246-272.
13. Lasiecka, I. and R. Triggiani: 1983, 'Dirichlet boundary control problem for parabolic equations with quadratic cost: Analyticity and Riccati feedback synthesis', SIAM J. Control Optimiz., 21, 41-67.

14. Lions, J.L. and E. Magenes: 1972, Non-Homogeneous Boundary Value Problems and Applications, I, II, Springer-Verlag, Berlin.
15. Triggiani, R.: 1980, 'Well posedness and regularity of boundary feedback parabolic systems', J. Diff. Eqns., 36, 347-362.
16. Zabczyk, J.: 1978, 'On decomposition of generators', SIAM J. Control Optimiz., 16, 523-534.
17. Yosida, K.: 1978, Functional Analysis, Springer-Verlag, Berlin.

P. Kotelenez

ON THE SEMIGROUP APPROACH TO STOCHASTIC EVOLUTION EQUATIONS

ABSTRACT

This paper deals with properties of mild solutions of stochastic partial differential equations (with and without delays). Both the case of a fixed Hilbert state space and the case of a nuclear Gel'fand triple as state space are analyzed. Examples of parabolic and hyperbolic equations are given.

Contents

L. Arnold and P. Kotelenez (eds.), Stochastic Space-Time Models and Limit Theorems, 95–139.

I. INTRODUCTION

This paper deals with properties of mild solutions of stochastic evolution equations of the type:

$$\left.\begin{aligned} dX(t) &= [AX(t) + \sum_{i=1}^{m} A_i X(t-i\rho)]dt \\ &\quad + dM(t) \qquad\qquad , \quad t\geq 0,\ \rho>0 \\ X(t) &= F(t) \qquad\qquad\qquad , \quad t\in[-m\rho,0] \end{aligned}\right\} \quad (1.1)$$

where A generates a strongly continuous semigroup $U(t)$ $(A\sim U(t))$ on a separable Hilbert space H and the A_i are relatively bounded perturbations of A. Here, $X_t(s) := X(t+s)$ for $s \in [-m\rho,0]$, $F(t)$ is a given initial process, and M is a certain noise process. Except for the last part of Section 3.1 we shall assume that M is state independent, i.e. (1.1) is linear.

For state dependent noise on H we refer to Kotelenez [32], Da Prato [12], and Tudor [50]. If $A_i \equiv 0$, the coefficients in $M(t,x)$ are essentially bounded operators and $u(t) \in G(t,\beta_T)$ (cf. the following Def. 1.1) then (1.1) has a unique mild solution under the same local Lipschitz and linear growth assumptions as in the case of ordinary stochastic equations (cf. Kotelenez [32] - $A = A(t)$ may even depend on t if $A(t)$ generates an evolution operator $U(t,s)$ satisfying a natural extension of Def. 1.1 and if the domain of $A(t)$ $Dom(A(t)))$ is independent of t). If $A_i \equiv 0$ and $U(t)$ analytic, bilinear equations with unbounded diffusion coefficients have been solved by DaPrato [12]. For explicit representations of solutions of bilinear equations -cf. Balakrishnan [4] and DaPrato Iannelli and Tubaro [14]. For bounded A_i, bounded noise coefficients and an H-valued Wiener process as driving term (1.1) has been investigated in Tudor [50 Moreover, for $A_i \equiv 0$ the Markov property for (1.1) was proved in Arnold, Curtain and Kotelenez [2] unde "natural" assumptions and for a special case the generator was computed in Curtain [10]. However, the assumption common to all the results mentioned above is that $M(t,x)$ is σ-additive on H.
The finitely additive case was

investigated by Dawson [17], [18] and Funaki [19] (cf. our Section 3.1).

Let us now assume $M(t,x) = M(t)$ state independent and look at the case $M(t)$ σ-additive and $M(t)$ finitely additive on H separately.

If $M(t)$ σ-additive on H (1.1) is an equation only on H and we obtain a "nice" mild solution if $U(t)$ belongs to the class of the following definition:

Definition 1.1

A strongly continuous semigroup $V(t)$ on a separable Hilbert space H with Hilbert space norm $|\cdot|_H$ is called contraction-type or, equivalently, to be an element of $G(1,\beta_T)$ if for any $T > 0$ there is a $\beta_T < \infty$ and an equivalent Hilbert space norm $\tilde{|}\cdot\tilde{|}_{H,T}$ s.t.

$$\tilde{|}V(t)\tilde{|}_{L(H)} \le e^{\beta_T t} \quad \text{for} \quad 0 \le t \le T ,$$

where the operator norm $\tilde{|}\cdot\tilde{|}_{L(H)}$ is computed w.r.t. $|\cdot|_{H,T}$.

Remark 1.1

1. Examples where the semigroup associated with a finite dimensional delay equation is in $G(1,\beta_T)$ are given in Banks and Kappel [5].
2. Another example is given in Lemma 2.1.2 for $U(t)$ analytic generated by A with $\mathrm{Dom}(A) = \mathrm{Dom}(A^*)$ where A^* is the adjoint operator.
3. Although $G(1,\beta_T)$ is large there are examples of Hilbert spaces and semigroups thereon which are not in $G(1,\beta_T)$ (cf. Chernoff [8])

Chapter 2 is devoted to stochastic evolution equations on the fixed Hilbert space H. Section 2.1 contains maximal inequalities for the stochastic convolution integral $\int_0^{\cdot} G(\cdot - s)dM(s)$ (which we also call stochastic *-integral), where $G(t)$ is the fundamental solution generated by the deterministic part of (1.1), and it contains regularity properties following from those inequalities. Moreover, some regularity results of DaPrato [12] are slightly generalized and it is shown that his assumption for maximal regularity of $\int_0^{\cdot} U(\cdot - s)dM(s)$ implies the coercivity of the generator A of $U(t)$. Section 2.2 gives a weak convergence criterion for (1.1).

In Chapter 3 the results are applied to stochastic space-time models where $M(t)$ is only finitely additive on H, e.g., $M(t)$ cylindrical Brownian motion on H. Consequently, we must redefine (1.1) on an enlarged space of distributions; it is shown that this can be done rigorously for a large class of linear equations. The "correct" state space for those equations is the strong dual Φ' of a nuclear Fréchet space and the solutions live on some Hilbert distribution subspaces of Φ'. For parabolic equations the case of a bounded domain D in R^n and the whole $R^n (n \geq 1)$ are studied separately. In the case of the bounded domain we can describe the smoothest possible state spaces for the solutions of our equations. Moreover, nonlinear equations on a bounded domain are also discussed. Finally, we define for certain second order equations a corresponding nuclear set-up. In all cases we give examples from the literature where these (linear) processes arise.

Notation and formal assumptions

All stochastic objects are defined on a complete stochastic basis $(\Omega, F, F_{t \geq 0}, P)$ with right continuous filtration, and all random variables depending on t are adapted to F_t. Hilbert spaces will usually be denoted by $(H, \langle\cdot,\cdot\rangle_H)$ where $\langle\cdot,\cdot\rangle_H$ is the scalar product on H. $L(H,K)$ is the set of bounded operators from H to K, $L(H) = L(H,H)$ and $|\cdot|_{L(H)}$ the operator norm.

For stability and invariant measures for (1.1) we refer to Ichikawa [26] (also for more references) and for maximum likelihood estimates to Bagchi and Borkar [3].

2. STOCHASTIC EVOLUTION EQUATIONS

2.1 Maximal Inequalities and Regularity for the Stochastic *-Integral

The first of our three basic maximal inequalities is a special case of an inequality proved in Kotelenez [33] and generalizes the stopped Doob inequality which Metivier and Pellaumail [40] obtained for martingales (cf. also Metivier [39]). Let c_0 be the universal constant in the Burkholder-Davis-Gundy inequality appearing in the upper estimate for the sup of a real valued martingale by the square root of its quadratic variation (Jacod [29]), and set

$$c_1 := 2(1+c_0^2)$$

$$\hat{\tau} := \begin{cases} \text{ess sup } \tau(\omega) , & \text{if it exists} \\ \infty & \text{otherwise} \end{cases} ,$$

where τ is a stopping time. Let us make the following assumption

A.2.1.1)

(i) M is an H-valued square integrable (s.i) cadlag martingale, $\langle M \rangle$ is its Meyer process and $[M^j]$ is the quadratic variation of its pure jump part (i.e., $M-M^j$ can have only totally inaccessible jumps);

(ii) for any $T > 0$ $U(t) \in G(1,\beta_T)$ with generator A .

Theorem 2.1.1

Under assumption A.2.1.1):

(I) $$\int_0^{\cdot} U(\cdot-s)dM(s) \in \begin{cases} \text{continuous a.s. if } M \text{ is continuous} \\ \text{cadlag a.s. if } M \text{ is cadlag} \end{cases}$$

(II) For any stopping time τ ,

$$E \sup_{0\le t<\tau} \left|\int_0^t U(t-s)dM(s)\right|_H^2 \le c_1 e^{4\beta_{\hat{\tau}}\hat{\tau}} E\{(\langle M\rangle+[M^j])(\tau-)\} \tag{2.1.1}$$

Remark 2.1.1

(i) Tubaro [49] has independently proved a similar result to (2.1.1)

(ii) (2.1.1) obviously implies the weaker inequality: For any $T > 0$ and $\delta > 0$,

$$P\{ \sup_{0\le t\le T} |\int_0^t U(t-s)dM(s)|_{H_1}^2 \ge \delta\} \le c_1 \frac{e^{4\beta_T T}}{\delta^2} E|M(T)|_H^2 \qquad (2.1.2)$$

which has been proved in Kotelenez [30] for evolution operators $U(t,s)$ (cf. Curtain and Pritchard [11] and Tanabe [47] for the definition and properties of evolution operators) instead of semigroups if $U(t,s)$ satisfies an estimate as in Def. 1.1 but without assuming the existence of a quasi-generator $A(t)$.

(2.1.2) has been generalized by Tudor [50] to cover stochastic delay-differential equations. In view of the application in Theorem 2.1.2 we shall give the basic maximal inequality with operators slightly more general than those considered by Tudor [50]. Therefore we make assumption

<u>A.2.1.2.0)</u>

(i) M is an H-valued s.i. cadlag martingale;

(ii) $K(t,s)$, $\tilde{K}(t,s)$, $0\le s\le t$, and $L(t,s,r)$, $0\le r\le s\le t$ are families of bounded linear operators on H which are strongly continuous in t and in s (resp. (s,r)) either strongly continuous or finitely valued such that for any $T > 0$ there is a constant $\beta_T < \infty$ with:

$$|K(t,s)|_{L(H)} \le e^{\beta_T(t-s)} \qquad 0 \le s \le t \le T$$

$$|L(t,s,r)|_{L(H)} \le \beta_T(t-s) \qquad 0 \le r \le s \le t \le T$$

$K(s,s) = I$ (identity operator) for all $s\ge 0$

$$\tilde{K}(t,r) = K(t,s)\tilde{K}(s,r) + L(t,s,r), \quad 0\le r\le s\le t .$$

From Chojnowska-Michalik [9] we have that $\int_0^\cdot \tilde{K}(\cdot,s)dM(s)$ admits a progressively measurable version, and by Gihman and Skorohod [21] we may assume that $|\int_0^\cdot \tilde{K}(\cdot,s)dM(s)|_H$ is separable.

<u>Lemma 2.1.1</u>

Under assumption A.2.1.2.0) for any $T > 0$ and all $\delta > 0$ we have

$$P\{ \sup_{0\le t\le T} |\int_0^t \tilde{K}(t,s)dM(s)|_H \ge \delta\} \le \frac{K(T)}{\delta^2} E|M(T)|_H^2, \quad (2.1.3)$$

with $K(T) := e^{4\beta_T T} \max\{(\overline{K}(T))^2, \beta_T \cdot K(T), \beta_T^2\}$

and $\overline{K}(T) := \sup_{0\le s\le t\le T} |\tilde{K}(t,s)|_{L(H)}$.

The proof is the same as the proof of Theorem 1 in Tudor [50] (cf. Kotelenez [30], Th. 1).

□

Let ρ be a fixed constant from $(0,\infty)$, $A, A_1, \ldots A_m$ linear operators on H, and let us consider the unperturbed delay-differential equation on H:

$$\frac{dX(t)}{dt} = AX(t) + \sum_{i=1}^{m} A_i X(t-i\rho) \quad (2.1.4)$$

We make the following assumption on (2.1.4):

A.2.1.2)
$A, A_1, \ldots, A_m$ are closed linear operators with dense domains in H s.t.

(i) $A \sim U(t) \in G(1,\beta_T)$ for all $T > 0$;

(ii) for each $i\in\{1,\ldots,m\}$ there is a function $c_i(\cdot)\in L_1[0,\rho]$ with
$|U(t)A_i h|_H \le c_i(t)|h|_H$ for all $h\in D(A_i)$, $i=1,\ldots,m$ and a.a. $t\in[0,\rho]$.

Obviously, (ii) in A.2.1.2) implies that for each i and a.a. $t\in[0,\rho]$ $U(t)A_i$ can be extended to a bounded operator on H, which we shall also denote by $U(t)A_i$ and

$$|U(t)A_i|_{L(H)} \le c_i(t).$$

Moreover, A.2.1.2) implies by Nakagiri [42] the existence of a fundamental solution $G(t)$ for (2.1.4), $t \ge 0$, with the representation

$$G(t) = \sum_{j=1}^{\infty} U_j(t-(j-1)\rho) \quad (2.1.5)$$

where

$$\left.\begin{array}{ll} U_j(t) \equiv 0 & \text{if } t < 0 \text{ for all } j \\ U_1(t) = U(t) & \text{if } t \geq 0 \\ U_j(t) = \sum_{i=1}^{(j-1)\wedge m} \int_0^t U(t-s)A_i U_{j-i}(s)ds & \\ \qquad \text{if } t \geq 0,\ j=2,3,.. & \end{array}\right\} \quad (2.1.6)$$

with "$\wedge$" denoting "min".

From the representation we see that

$G(t)$ is strongly continuous, bounded, and $G(0) = I$,

and for $0 \leq r \leq s \leq t$

$$G(t-r) = U(t-s)G(s-r) + R(t,s,r) \quad (2.1.7)$$

with

$$R(t,s,r) := \sum_{k=2}^{\infty} \sum_{i=1}^{(k-1)\wedge m} \int_{s-r-(k-1)\rho}^{t-r-(k-1)\rho} U(t-r-(i-1)\rho-u) A_i U_{k-i}(u)du .$$

Thus, $G(t)$, $U(t)$ and $R(t,s,r)$ satisfy assumption A.2.1.2.0), (ii), on $K(t,s)$, $K(t,s)$ and $L(t,s,r)$. On the other hand, setting for $T > 0$

$$s(p) := \frac{Tj}{2^p} \quad \text{if} \quad s \in \left(\frac{Tj}{2^p}, \frac{T(j+1)}{2^p}\right),\ j=0,1,...,2^p-1$$

we easily see that

$$\int_0^{\cdot} G(\cdot-s(p))dM(s) \text{ is } \begin{cases} \text{continuous if } M \text{ is continuous} \\ \text{cadlag if } M \text{ is cadlag} \end{cases}$$

and for $p \geq q$

$$\tilde{K}_q^p(t,s) := G(t-s(p)) - G(t-s(q))$$
$$K(t,s) := U(t-s)$$
$$L_q^p(t,s,r) := R(t,s,r(p)) - R(t,s,r(q))$$

also satisfy A.2.1.2.0), ii). As a consequence we obtain (cf. Kotelenez [30], Th. 2):

Theorem 2.1.2
Under assumption A.2.1.2):

(I) $\int_0^{\cdot} G(\cdot-s)dM(s)$ is $\begin{cases} \text{continuous a.s. if } M \text{ is continuous a.s.} \\ \text{cadlag a.s. if } M \text{ is cadlag a.s.} \end{cases}$

(II) For any $T > 0$ there is a constant $K(T) < \infty$ s.t. for all $\delta > 0$

$$P\{\sup_{0\le t\le T} |\int_0^t G(t-s)dM(s)|_H \ge \delta\} \le \frac{K(T)}{\delta^2} |M(T)|_H^2 . \quad (2.1.8)$$

where $K(T)$ is a function of β_T and the $c_i(\cdot)$ in A.2.1.2) as described in Lemma 2.1.1.

We now want to derive a maximal inequality for the stochastic *-integral w.r.t. a norm stronger than the norm of the state space of the noise process M. In order to obtain this result for all $U(t)$ which are generated by self-adjoint closures A of strongly elliptic operators on a bounded domain (and for other operators on R^n - cf. Remark 3.2.4) we somewhat generalize Dawson's [17] assumptions on the spectrum of A.

A.2.1.3)

(i) M is an H-valued continuous martingale with tensor quadratic variation $|M|$ (cf. Metivier and Pellaumail [41]) s.t. for all $T > 0$ and $\eta > 0$ there is a nuclear positive self-adjoint operator Q_η^T on H with:

$$P\{\sup_{0\le s\le t\le T} [M](t) - [M](s) \le Q_\eta^T(t-s)\} \ge 1 - \eta$$

where for two self-adjoint nonnegative definite operators A,B on H : $A \le B \Leftrightarrow B-A$ nonnegative definite;

(ii) $A \sim U(t)$, is self-adjoint and has a discrete spectrum $\{\lambda_n\}$ s.t. there is a $\gamma_0 < 0$ with

$$\Sigma(\beta-\lambda_n)^\gamma < \infty \text{ for all } \gamma < \gamma_0.$$

where $\beta \ge 0$ is chosen s.t. $(A-\beta)$ is negative definite.

(iii) $A_1,\ldots,A_m$ as in A.2.1.2), and, moreover, all A_i commute with A .

Remark 2.1.2

(i) Condition (i) in A.2.1.3) is satisfied, if, e.g., $M(t) = \int_0^t \Phi dW(s)$ for $\Phi \in \Lambda^2(K,H,\mathcal{O},W)$, where W is a K-valued Wiener process with covariance operator Q . In this case,

$$[\![M]\!](t) = \Phi Q \Phi^*$$

and the existence of Q^T in A.2.1.3) follows from the definition of $\Lambda^2(K,H,\mathcal{O},W)$ (cf. [39])

(ii) Condition (ii) in A.2.1.3) implies the coercivity of $-A$, whence $U(t) \in \mathcal{G}(1,\beta)$ and $U(t)$ analytic. Indeed, since $A \sim U(t)$ there is a $\beta \geq 0$ s.t. $\lambda_n + \beta > 0$ for all n . Set

$$V := \mathrm{Dom}((-A+\beta)^{\frac{1}{2}}), <\cdot,\cdot>_V := <(-A+\beta)^{\frac{1}{2}}\cdot,(-A+\beta)^{\frac{1}{2}}\cdot>_H .$$

Then for $\varphi \in \mathrm{Dom}(A)$

$$<-A\varphi,\varphi>_H = <\varphi,\varphi>_V - \beta<\varphi,\varphi>_H .$$

(iii) Condition (i) implies that M has Hölder modulus $\mu = \frac{1}{2}$ in t (where the Hölder modulus is the sup over all possible Hölder exponents - cf. Kotelenez [32]) whence from the analyticity of $U(t)$ and the work of DaPrato, Iannelli and Tubaro [15]

$$\int_0^{\cdot} U(\cdot - s)dM(s) \text{ has Hölder modulus } \mu = \tfrac{1}{2} \text{ (in } t\text{) on } H.$$

Let us define

$$(H_\alpha, <\cdot,\cdot>_\alpha) := (\mathrm{Dom}((-A+\beta)^{\frac{\alpha}{2}}), <(-A+\beta)^{\frac{\alpha}{2}}\cdot,(-A+\beta)^{\frac{\alpha}{2}}>_H), \alpha \geq$$

where "Dom" denotes the domain of the operator. Hence

$$H = H_0 \,,\quad V = H_1 \,.$$

Let "v" denote "max" . In the following theorem $G(t)$ is the fundamental solution of (2.1.4) with representation (2.1.5).

Theorem 2.1.3

Under assumption A.2.1.3):

(I) For any $T > 0$, all $\mu \in [0,\frac{1}{2})$ and all $\alpha \in [0,1)$

$$\int_0^{\cdot} G(\cdot-s)dM(s) \in C^{\mu}([0,T];H_0) \cap C([0,T];\ H_\alpha) \quad \text{a.s.}$$

(II) For all $T > 0$ and all $\alpha \in [0,1)$ and all $r > \frac{2|\gamma_0|+1}{1-\alpha} + \frac{1}{2}$ there is a constant $K = K(T,\alpha,r,\beta,c_i,i=1,\ldots,m) < \infty$ such that for any $\eta,\delta > 0$'

$$P\{\sup_{0\le t\le T} |\int_0^t G(t-s)dM(s)|_\alpha \ge \delta\} \le \frac{K}{\delta^{2r}} \sum_n \lambda_n^{\gamma}[TrQ_n^T]^r + \eta \tag{2.1.9}$$

with $\gamma := \dfrac{(r-\frac{1}{2})(1-\alpha)-1}{2}$.

Proof

(i) Let $b(t)$ be a real valued standard Brownian motion and λ and a some numbers > 0 . The following estimate is due to Dawson [17] and basic for the proof of our theorem:

There is a constant $K < \infty$ s.t. for all sufficiently large $a\lambda$ and for any $T > 0$:

$$P\{\sup_{0\le t\le T} |\int_0^t e^{-\lambda(t-s)}db(s)| \ge a^{1/2}\}$$

$$\le K\{(a\lambda)^{\frac{1}{2}}\exp(-a\lambda)[\lambda T+\log((a\lambda)^{\frac{1}{2}})] \tag{2.1.10}$$

$$+(a\lambda)^{\frac{1}{4}}\exp(-\frac{a\lambda}{2})(\log((a\lambda)^{\frac{1}{2}}\exp(-2\lambda T)\}.$$

(ii) Since we may consider $\tilde{\lambda}_n := \lambda_n + \beta$ instead of λ_n , we may assume w.l.o.g. $0 < \lambda_1 \le \lambda_2 \le \ldots$, and that there is a CONS $\{\varphi_n\}$ of eigenvectors of A with λ_n as eigenvalues. Then, we easily see that $\varphi_n^\alpha := \lambda_n^{-\frac{\alpha}{2}}\varphi_n$, $n \in \mathbb{N}$, is a CONS for H_α , $\alpha \ge 0$, and for any $x \in H_0$ we set

$$\langle x, \varphi_n^{\alpha} \rangle_{\alpha} := \langle x, \varphi_n \rangle_0 \, \lambda_n^{\frac{\alpha}{2}} .$$

(iii) Fix $\alpha \in [0,1)$ and set $C(\eta, T, \mu) = [\mathrm{Tr}(Q_{\eta}^{T})]^{-1}$ Denote by μ_n the eigenvalues of Q_{η}^{T} and by $[M_n](s)$ the quadratic variation of $\langle M(s), \varphi_n \rangle_0$. Then for $T > 0$, $\delta > 0$:

$$P\{ \sup_{0 \le t \le T} | \int_0^t U(t-s) dM(s) |_{\alpha}^{2} \ge \delta^2 \}$$

$$\le \sum_n P\{ \sup_{0 \le t \le T} (\int_0^t e^{-\lambda_n (t-s)} \langle dM(s), \varphi_n^{\alpha} \rangle_{\alpha})^2 \ge \delta^2 C(\eta, T, \mu) \mu_n \}$$

$$= \sum_n P\{ \sup_{0 \le t \le T} (\int_0^t e^{-\lambda_n (t-s)} \langle dM(s), \varphi_n \rangle_0)^2 \ge \delta^2 C(\eta, T, \mu) \mu_n \lambda_n^{-}$$

$$= \sum_n P\{ \sup_{0 \le t \le T} (\int_0^t e^{-\lambda_n (t-s)} db([M_n](s)))^2 \ge \delta^2 C(\eta, T, \mu) \mu_n \lambda_n^{-}$$

(since $\langle M(s), \varphi_u \rangle$ is a $[M_u](s)$ - time changed standard Brownian motion $b(s)$)

$$\le \sum_n P\{ \sup_{0 \le t \le T} (\int_0^t e^{-\lambda_n (-s)} db(\mu_n s))^2 \ge \delta^2 C(\eta, T, \mu) \mu_n \lambda_n^{-\alpha} \}$$

by A.2.1.3)(i)

$$= \sum_n P\{ \sup_{0 \le t \le T} (\int_0^t e^{-\lambda_n (t-s)} db(s))^2 \ge \delta^2 C(\eta, T, \mu) \lambda_n^{-\alpha} \}$$

$$\le \sum_n K\{ ([a_n \lambda_n]^{\frac{1}{2}} \exp(-a_n \lambda_n) [\lambda_n T + \log((a_n \lambda_n)^{\frac{1}{2}})]$$

$$+ (a_n \lambda_n)^{\frac{1}{4}} \exp(\frac{-a_n \lambda_n}{2}) (\log((a_n \lambda_n)^{\frac{1}{2}}))^{\frac{1}{2}} \exp(-2\lambda_n T)$$

by (2.2.10) with $a_n := \delta^2 C(\eta, T, \mu) \lambda_u^{-\alpha}$ since $a_n \lambda_n \sim \lambda_n^{1-\alpha} \to \infty$, whence for any $r > 0$ there is a $C(K, r)$ s.t. the r.h.s. of the last inequality can be estimated from above by

$$C(K,r)\sum_n\{(a_n\lambda_n)^{\frac{1}{2}}(a_n\lambda_n)^{-r}[\lambda_n T+\log((a_n\lambda_n)^{\frac{1}{2}})]$$

$$+\ (a_n\lambda_n)^{\frac{1}{4}}(a_n\lambda_n)^{-\frac{r}{2}}(\log((a_n\lambda_n)^{\frac{1}{2}})^{\frac{1}{2}}\}\ .$$

If we now choose $r > \frac{2|\gamma_0|+1}{1-\alpha} + \frac{1}{2}$ we obtain (2.1.9) with $U(t)$ instead of $G(t)$. Moreover, by A.2.1.3)(i) we obtain (2.1.9) for $\int_0^t U(t-s)dM(s+r)$ uniformly in r from an arbitrary fixed interval $[0,\tilde{T}]$.

(iv) First note that $\mathcal{B} \in L(H_0)$ and the fact that B commutes with A implies that B also commutes with $(\beta-A)^{\frac{\alpha}{2}}$ (and with $U(t)$), whence B restricted to H_α is in $L(H_\alpha)$ with $||B||_{L(H_\alpha)} = ||B||_{L(H_0)}$ (cf. Section 3.1).

From (2.1.7) we obtain

$$\int_0^t G(t-s)dM(s) = \int_0^t U(t-s)dM(s) + \int_0^t R(t,s,s)dM(s)$$

and

$$\int_0^t R(t,s,s)dM(s)$$

$$
\left.\begin{aligned}
&= \sum_{k=2}^{\infty} \sum_{i=1}^{(k-1)\wedge m} \int_0^t \\
&\quad \left[\int_0^{t-s-(k-1)\rho} U(t-s-(k-1)\rho-u)A_i U_{k-i}(u)du\right] dM(s) \\
&= \sum_{k=2}^{\infty} \sum_{i=1}^{(k-1)\wedge m} \int_0^t A_i U_{k-i}(u) \\
&\quad \left[\int_0^{t-u-(k-1)\rho} U(t-u-(k-1)\rho-s)dM(s)\right] du
\end{aligned}\right\} \quad (*)
$$

by the stochastic Fubini theorem (Chojnowska-Michalik [9]) and since by assumption A_i and, consequently, also $A_i U_{k-i}(u)$ commute with A and, therefore, with $U(t)$.

On the other hand, A.2.1.2) implies by Nakagiri [42] for any $T > 0$ and $i=1,\ldots,(k-1)\wedge m$, $k = 2,3,\ldots$ the existence of $\tilde{C}_{i,k,T}(\cdot) \in L_1[0,T]$ s.t.

$$|A_i U_{k-i}(u)|_{L(H_0)} \le \tilde{C}_{i,k,T}(u) \text{ for a.a. } u \in [0,T] .$$

But the l.h.s. in the last inequality is equal to $|A_i U_{k-i}(u)|_{L(H_\alpha)}$. Therefore,

$$\sup_{0\le t\le T} \left|\int_0^t R(t,s,s)dM(s)\right|_\alpha$$

$$\le \sum_{k=2}^{\tilde{T}} \sum_{i=1}^{(k-1)\wedge T} \int_0^T \tilde{c}_{i,k,T}(u) \cdot \sup_{0\le t\le T} \left|\int_0^t U(t-s)dM(s+u+(k-1)\rho)\right|$$

with $\tilde{T}$ the smallest natural number $\ge \frac{T}{\rho} + 1$, since for larger k the integrands in (*) become 0. Consequently, by step (iii) we obtain (II).

(v) The existence of the spatially smooth version for $\int_0^{\cdot} G(\cdot-s)dM(s)$ is now an easy consequence by the Borel-Cantelli theorem. By Remark 2.1.2,(iii), $\int_0^{\cdot} U(\cdot-s)dM(s)$ is in $C^{\mu}([0,T];H_0)$ a.s. for all $\mu \in [0,\frac{1}{2})$ which implies by the commutativity

assumption w.r.t. A_i, A that
$\int_0^{\cdot} R(\cdot,s,s)dM(s) \in C^{\mu}([0,T];H_0)$, whence we obtain (I).

□

In [12] DaPrato has obtained spatial regularity results for the stochastic *-integral by semigroup methods if the semigroup is analytic. These results were applied by DaPrato to linear SPDE . Such regularity results had been obtained earlier by Pardoux [43] and Krylov and Rozovskij [35] via a variational approach under a coercitivity assumption on the generator A (and a monotonicity assumption if A is nonlinear). In this last part of Section 2.1 we shall extend DaPrato's maximal regularity results to more general noise terms and to the fundamental solution $G(t)$ of (2.1.5) assuming in this latter case that $U(t)$ is analytic. Accordingly, we assume all Hilbert spaces to be complex for this last part of Section 2.1. DaPrato's assumption for maximal regularity is the following:

A.2.1.4)
$A \sim U(t)$ is analytic, and $\mathrm{Dom}(A) = \mathrm{Dom}(A^*)$,
where A^* is the adjoint operator of A .

Before proceeding further we want to clarify the relation of assumption A.2.1.4) to the coercivity assumption which is basic to the variational approach (cf. Lions and Magencs [38] and Pardoux [43]):

Lemma 2.1.2
(i) If $-A$ defines a coercive quadratic form, then $U(t)$ is analytic, and $U(t) \in G(1,\beta)$
(ii) Under assumption A.2.1.4) $-A$ defines a coereive form w.r.t. an equivalent scalar procut $\langle\langle\cdot,\cdot\rangle\rangle_0$ on H .

Proof
(i) follows from Tanabe [47], Section 3.6 (A is called regularly dissipative in [47]).
(ii) DaPrato and Grisvard [13] have shown that under assumption A.2.1.4) there is an extension $(\hat{A},\hat{U}(t))$ of $(A,U(t))$ onto a Hilbert space $(\hat{H},(\cdot,\cdot))$ $(\hat{H},(\cdot,\cdot)) \supset (H,\langle\cdot,\cdot\rangle$ (in the sende of continuous inclusion) s.t. $H = \mathrm{Dom}(\hat{A})$, $\hat{U}(t)$ analytic, and

$|\hat{U}(t)|_{L(\hat{H})} = |U(t)|_{L(H)}$. Let us denote by

$$(H_{-\varepsilon}, (\cdot,\cdot)_{-\varepsilon}) \;, \quad \varepsilon \in (0,2)$$

the interpolation spaces between H and $\hat{H}$ in the sense of Butzer and Berens [7] (cf. DaPrato [12]) which are equal to the interpolation spaces of Lions and Magenes [38], $[H,\hat{H}]\varepsilon/2$ (cf. Butzer and Berens [7], Theorems 3.4.2 and 3.5.3). By interpolation we obtain that $(\hat{A},\hat{U}(t))$ has a restriction $(A_{-\varepsilon},U_{-\varepsilon}(t))$ on $H_{-\varepsilon}$ s.t. $U_{-\varepsilon}(t)$ is analytic (cf. Davies [16], Th. 2.39 and Lions and Magenes [38], Ch. 1, Section 5.1). Now we fix $\varepsilon \in (0,1)$ and interpolate between $H_{-\varepsilon}$ and $D(A_{-\varepsilon})$ using the method of Butzer and Berens (assuming w.l.o.g. that the spectrum of $A \subset (-\infty,0)$):

$$(H_{-\varepsilon+\gamma}, \langle\langle\cdot,\cdot\rangle\rangle_{-\varepsilon+\gamma}) \;, \quad \gamma \in (0,2)$$

is the subspace of those $x \in H_{-\varepsilon}$ for which $\langle\langle x,x\rangle\rangle_{-\varepsilon+\gamma} =: ||x||^2_{-\varepsilon+\gamma} < \infty$, where

$$\langle\langle x,y\rangle\rangle_{-\varepsilon+\gamma} := \int_0^\infty t^{1-\gamma}(A_{-\varepsilon}U_{-\varepsilon}(t)x, A_{-\varepsilon}U_{-\varepsilon}(t)Y)_{-\varepsilon}dt. \tag{2.1.11}$$

Note that

$$H_0 = H$$

with equivalent norms. Then set

$$V := H_1 = [D(A),H_0]_{\frac{1}{2}}$$

and after identifying H_0 with its dual H_0^* we obtain

$$V^* = H_{-1} = [H_0,\hat{H}]_{\frac{1}{2}}$$

from Lions and Magenes [38], Ch. 1, Th. 6.2. Moreover, we have

$$A \in L(D(A),H_0) \quad \text{and} \quad \hat{A} \in L(H_0,\hat{H})$$

which implies by Lions and Magenes [38], Ch. 1,

Th. 5.1, that

$$A_{-1} \in L(V,V^*) .$$

Hence, denoting by $(\cdot,\cdot)_{V^*,V}$ the duality between V^* and V, we obtain for some $c > 0$:

$$|(A_{-1}x,y)_{V^*,V}| \le c|x|_1 \cdot |y|_1$$

for $x,y \in H_1 = V$.

On the other hand, if $x \in D(A)$ then (for $\varepsilon \in (0,1)$)

$$\begin{aligned}(A_{-1}x,x) &= \langle\langle Ax,x\rangle\rangle_0 \\ &= \int_0^\infty t^{1-\varepsilon}(A^2U(t)x, AU(t)x)_{-\varepsilon}\,dt \\ &= \int_0^\infty t^{1-\varepsilon}(A\tfrac{\partial}{\partial t}U(t)x,\ AU(t)x)_{-\varepsilon}dt \\ &= -\int_0^\infty t^{1-\varepsilon}(AU(t)x,\ A\tfrac{\partial}{\partial t}U(t))_{-\varepsilon}dt \\ &\quad -(1-\varepsilon)\int_0^\infty t^{-\varepsilon}|AU(t)x|_{-\varepsilon}^2 dt\end{aligned}$$

by partial integration, whence

$$\mathrm{Re}(A_{-1}x,x)_{V^*,V} \le \frac{-(1-\varepsilon)}{2}||x||_1^2$$

for $x \in D(A)$ and, consequently, by continuity, also for $x \in H_1 = V$.

□

<u>Corollary 2.1.1</u>

If assumption A.2.1.4) holds, then $U(t) \in G(1,\beta)$.

Instead of A.2.1.4) we may by Lemma 2.1.2 make the following coercivity assumption on A in (2.1.4)

<u>A.2.1.5</u>

V and H are separable Hilbert spaces with

$$V \to H$$

and the topology of V is stronger than that of H. There is a quadratic form $a(x,y)$ on V and constants $c, \mu, \beta > 0$ s.t.

$$|a(x,y)| \le c|x|_V \cdot |y|_V$$

$$\mathrm{Re}\, a(x,x) \ge \mu|x|_V^2 - \beta|x|_H^2$$

Remark 2.1.3

We take A to be the restriction to H of $\hat{A}$ which exists by the Lax-Milgram theorem through

$$(\tilde{A}x,y)_{V^*,V} = a(x,y)$$

where V^* is the dual of V. Then by Tanabe [47] both A and $\hat{A}$ generate analytic semigroups $U(t)$ and $\tilde{U}(t)$ on H and V^*, respectively. Moreover, $U(t) \in G(1,\beta)$.

In the following Theorem 2.1.4 G is the fundamental solution for (2.1.4) as given in (2.1.5).

Theorem 2.1.4

Let M be an H-valued square integrable martingale. A is defined in Remark 2.1.3 under assumption A.2.1.5). For $A_1,\dots,A_m$ assume A.2.1.2)., (ii) and A.2.1.3), (ii).

Then

$$\int_0^{\cdot} G(\cdot-s)\,dM(s) \in \begin{cases} D([0,\infty);H) \text{ a.s. if } M \text{ is cadlag a.s.} \\ C^{\mu}([0,T);H) \text{ a.s. if } M \in C^{\mu}([0,T];H) \\ \qquad\qquad \text{a.s., } \mu \in [0,\frac{1}{2}), T>0 \\ L_2^{F}([0,T]\times\Omega;V) \end{cases}$$

Proof

(i) Take $\varepsilon \in (0,\frac{1}{2})$ and set

$$H_{-\varepsilon} := [H,V^*]_{\frac{1}{2}} .$$

By interpolation we see that $\bar{U}$, the restriction of $\tilde{U}(t)$ onto $H_{-\varepsilon}$, is analytic (cf. proof of Lemma 2.1.2)

(ii) Let $[\bar{M}]$ denote the quadratic variation of M in $H_{-\varepsilon}$ and $\bar{Q}^M$ the "probabilistic" covariance operator of M on $H_{-\varepsilon}$ (i.e.; $\bar{Q}^M$ is the Radon-

Nikodym derivative of the tensor valued measure generated by the tensor quadratic variation over $H_{-\varepsilon}$ w.r.t. the Doleans measure generated by $[\overline{M}]$ - cf. Metivier and Pellaumail [41]). $\overline{\mathrm{Tr}}$ is the trace on $H_{-\varepsilon}$. Then, following DaPrato's proof in [12] (cf. also Metivier and Pellaumail, loc.cit.) we obtain (repeatedly using Fubini's theorem) for $\vartheta = \varepsilon + \frac{1}{2}$ that

$$E\int_0^T \left|\int_0^t U(t-s)dM(s)\right|_V^2 dt$$

$$= E\int_0^T\int_0^\infty\int_0^t h^{1-2\vartheta}\,\overline{\mathrm{Tr}}(\overline{A}\overline{U}(t-s+k)\overline{Q}_s^M\overline{U}^*(t-s+h)\overline{A}^*)d[\overline{M}](s)dhdt$$

$$= E\int_0^\infty h^{1-2\vartheta}\int_0^T\int_s^T \overline{\mathrm{Tr}}(\overline{A}\overline{U}(t-s+h)\overline{Q}_s^M\overline{U}^*(t-s+h)A^*)dtd[\overline{M}](s)dh$$

$$\leq E\int_0^\infty h^{1-2\vartheta}\int_0^\infty\int_0^T \overline{\mathrm{Tr}}(\overline{A}\overline{U}(p+h)\overline{Q}_s^M\overline{U}^*(p+h)\overline{A}^*)d[\overline{M}](s)dpdh$$

by change of variable $t = p + s$

$$= E\int_0^\infty\int_0^r (r-p)^{1-2\vartheta}\int_0^T \overline{\mathrm{Tr}}(\overline{A}\overline{U}(r)\overline{Q}_s^M\overline{U}^*(r)\overline{A}^*)d[\overline{M}](s)dpdr$$

by change of variable $h = r - p$

$$= \frac{1}{1-2\varepsilon}\,E\int_0^\infty r^{1-2\varepsilon}\int_0^T \overline{\mathrm{Tr}}(\overline{A}\overline{U}(r)\overline{Q}_s^M\overline{U}^*(r)A^*)d[\overline{M}](s)dr$$

$$= \frac{1}{1-2\varepsilon}\,E||M(T)||_0^2 .$$

(iii) From this we obtain as in step (iv) of the proof of Th. 2.1.3 that $\int_0^{\cdot} G(\cdot-s)dM(s) \in L_2^F([0,T]\times\Omega;V)$.

(iv) The other statements ar obvious in view of Th. 2.1.1 Remark 2.1.2, (iii) and the proof of Th. 2.1.3.

□

Remark 2.1.4
DaPrato [12] has shown by an example that in general there cannot be found a smaller (smoother) Hilbert space $K \subsetneq V$ such that $\int_0^{\cdot} U\cdot-s)dM(s)$ lives in K if M is in H .

Remark 2.1.5
Let W be a K-valued Wiener process, σ an $L(K,H)$-valued adapted process, and $U(t)$ an analytic semigroup with generator A . Then DaPrato has shown in [12] (without the additional-coercivity-assumption $D(A) = D(A^*)$) that for all $\alpha \in [0,1)$ and $T > 0$:

(i) $\int_0^{\cdot} U(\cdot-s)\sigma(s)dW(s) \in L_2^{\mathcal{F}}([0,T]\times\Omega;\ H_\alpha)$

if $\sigma \in L_2^{\mathcal{F}}([0,T]\times\Omega; L(K,H))$;

(ii) $\int_0^{\cdot} U(\cdot-s)\sigma(s)dW(s) \in C([0,T];\ L^2(\Omega;H_\alpha)$

if $\sup_i |\sigma e_i|_{C([0,T];L_2(\Omega;H)}$

where $\{e_i\}$ are the eigenvectors of the covariance operator of W .
Clearly, we can easily obtain the generalizations of DaPrato's results analogous to Theorem 2.1.4 just with H_α $(\alpha\in[0,1))$ instead of $H_1 = V$ and without the coercivity assumption.

2.2 Weak Convergence for Linear Stochastic Evolution Equations

As an important application of the maximal inequalities of Section 2.1 we obtain weak convergence of solutions of linear stochastic evolution equations as a consequence of the weak convergence of their martingale integrators. E.g., on the basis of (2.1.2) the following was proved in Kotelenez [34]:

Assume $\{M^n\}_{n\in\mathbb{N}\cup\{0\}}$ are H-valued locally s.i. (l.s.i.) cadlag martingales s.t.

$$M'' \Rightarrow M^0 \quad \text{on} \quad D([0,\infty);H) \text{ (converges weakly)}$$

and $\{U^n(t,s)\}_{n\in N\cup\{0\}}$ are evolution operators in $G(1,\beta)$ (uniformly w.r.t. β) with quasi-generators $A^n(t)$ s.t.

$$A^n(t) \to A^o(t) \quad \text{suitably.}$$

Then

$$\int_0^{\cdot} U^n(\cdot,s)dM^n(s) \Rightarrow \int_0^{\cdot} U^o(\cdot,s)dM^o(s).$$

We can obtain similar results for linear stochastic evolution equations with delay on the basis of (2.1.8) or (2.1.9).

A.2.2.1)

(I) There are $(A_k, A_{i,k}, i=1,\dots,m)_{k\in N\cup\{0\}}$ satisfying A.2.1.2) with β_T and $c_i(t)$ uniformly in n s.t.

(i) $\Phi := \bigcap_{n,k=0}^{\infty}$ is a core for A_o, where A_k^n is the n-th power of A_k;

(ii) $Dom(A_{i,k}) \supset Dom(A_k)$ $i = 1,\dots,m, k \in N\cup\{0\}$;

(iii) $A_{i,k}$ commute with $A_k, i=1,\dots,m, k \in N\cup\{0\}$;

(iv) for all $\varphi \in \Phi$ $A_k\varphi \to A_o\varphi$, $A_{i,k}\varphi \to A_{i,o}\varphi$, $i=1,\dots,m$, as $k \to \infty$.

(II) There is a sequence of H-valued l.s.i. cadlag martingales M_k on complete stochastic bases $(\Omega^k, F^k, F_t^k, P^k)$ with right continuous filtration, $k \in N\cup\{0\}$,

s.t. $M_k \Rightarrow M_o$ on $D([0,\infty);H)$.

Theorem 2.2.1

Assume A.2.2.1) and let $G_k(t)$ be the fundamental solutions of (2.1.4) with A_k, $A_{i,u}$ instead of A, A_i.

Then

$$\int_0^{\cdot} G_k(\cdot - s)dM_k(s) \Rightarrow \int_0^{\cdot} G_0(\cdot - s)dM_0(s) \text{ on } D([0,\infty);H).$$

<u>Proof</u>

(i) Since $A_{i,k}$ commute with A_k we have on Φ that

$$G_k(t) = U_k(t) + \sum_{i=2}^{l} \sum_{j=1}^{i-1} \sum_{\substack{p_1+\dots+p_j = l \\ 1 \le p_i, \dots, p_j \le m}} \frac{1}{j!}(t-(i-1)\rho)^j U_k(t-(i-1)\rho) A_{p_1} \dots A_{p_j}$$

(Nakagiri [42], (3.1.4)).

(ii) Let $\{\varphi_n\}$ be a CONS for H with $\varphi_n \in \Phi$ for all n, denote by $L(\varphi_1, \dots, \varphi_{n_0}\}$ the linear span of $\varphi_1, \dots, \varphi_{n_0}$ and by Π_{n_0} the projection of H onto $L(\varphi_1, \dots, \varphi_{n_0})$ and set $\Pi_{n_0}^{\perp} := I - \Pi_{n_0}$ (I is the identity). Then

$$\int_0^t G_k(t-s)dM_k(s) = \int_0^t G_k(t-s)\Pi_{n_0}dM_k(s) + \int_0^t G_k(t-s)\Pi_{n_0}^{\perp}dM_k(s)$$

$$= \Pi_{n_0}M_k(t) + \int_0^t \frac{\partial}{\partial s}G_k(t-s)(\Pi_{n_0}M_k(s))ds + \int_0^t G_k(t-s)\Pi_{n_0}^{\perp}dM_k(s)$$

by partial integration.

Step (i) shows how to compute $\frac{\partial}{\partial s}G_k(t-s)(\Pi_{n_0}M_k(s))$ whence by our assumptions and the continuous mapping theorem (cf. Billingsley [6], Ch. I, Th. 5.1) for any n_o we have

$$\int_0^\cdot G_k(\cdot-s)\Pi_{n_o}dM_k(s) \Rightarrow \int_0^\cdot G_o(\cdot-s)\Pi_{n_o}dM_o(s)$$

on $D([0,\infty);H)$. The uniform negligibility of $\int_0^\cdot G_k(\cdot-s)\Pi_{n_o}^{\perp}dM_k$ follows from (2.1.8) exactly as in Th. 1.1 of Kotelenez [34].

□

It would be easy to give an analogous theorem on weak convergence on $D([0,\infty);H_\alpha)$, $\alpha \in [0,1)$ using (2.1.9).

3. STOCHASTIC SPACE-TIME MODELS

Here we shall consider nuclear spaces Φ^* as state spaces for spatially distributed systems. Φ^* is supposed to be the strong dual of a nuclear Fréchet space Φ (test function space). Since in this case Φ is the projective limit of a sequence of Hilbert spaces (cf. Schaefer [46]) we can anlyze our systems on a scale of Hilbert spaces

$$\Phi \subset \ldots \subset H_\alpha \subset \ldots \subset H = H^* \subset \ldots \subset H_{-\alpha} \subset \ldots \subset \Phi^* ,$$
$$\alpha \in N \quad \text{or} \quad \alpha \in R_+ , \tag{3.1}$$

with dense continous imbeddings s.t. for any $\alpha \in R$ (Z) there is a $\beta > \alpha$, $\beta \in R(Z)$, and

$$H_\beta \to H_\alpha$$

is nuclear. $H_{-\alpha}$ is the strong dual of H_α (cf. Gel'fand Vilenkin [20]). Typically $H_o = L_2(D)$, $D \subset R^n$, $n \in N$ (s. Sections 3.1-3.3). In order to apply the results of Sections 2.1, 2.2 to this case we must show that if we start with a "nice" semigroup $U(t)$ on H_o, we can extend $U(t)$ to a "nice" semigroup $U_{-\alpha}(t)$ on $H_{-\alpha}$ where $\alpha \geq 0$ depends on the properties of the system. (For the application of 2.3 - see Theorem 3.1.1).

The following definition is motivated by the fact that $U(t) \in G(1,\beta_T)$ was basic for our previous results. (In view of our applications we restrict ourselves to semigroups).

Definition 3.1

A semigroup $U(t) \in G(1,\beta_T)$ with generator A on H_0 is called $G(1,\beta)$-extendible (resp. restrictible onto $H_{-\alpha}$, $\alpha > 0$, (resp. H_α) if there is a $\beta(T,-\alpha)$ $(\beta(T,\alpha))$ s.t. $U(t)$ can be extended to a bounded operator $U_{-\alpha}(t)$ on $H_{-\alpha}$, $t \geq 0$, (restricted to $U_\alpha(t)$ on H_α) and $U_{-\alpha}(t) \in G(1,\beta(T,-\alpha))(U_\alpha(t) \in G(1,\beta(T,\alpha))$ with generator $A_{-\alpha}$ (A_α) which is an extension (restriction) of A.

Remark 3.1

1. A detailed analysis of the extendibility of analytic semigroups was carried through by DaPrato and Grisvard [13].
2. Clearly, the extensions of $U(t)$, if they exist, are unique.

Remark 3.2

1. (3.1) is in many applications (s. Sections 3.1 - 3.3 below) the natural state space of generalized Gauss processes. Gaussian martingales (independent increments!) were described by Itô [27] on the Schwartz set up

$$S \subset H_\alpha \subset L_2(R^n) \subset H_{-\alpha} \subset S' .$$

Now let us assume we have $U_{-\alpha}(t,s) \in G(1,\beta(T,-\alpha)$ on $H_{-\alpha}$ and a Gaussian martingale M on $H_{-\alpha}$. Then $Y(\cdot) := \int_0^{\cdot} U_{-\alpha}(\cdot,s)dM(s)$ is a continuous Gauss-Markov process on $H_{-\alpha}$ (by Th. 2.1.1 and Arnold, Curtain and Kotelenez [2]). On the other hand, assume the existence of a continuous Gauss-Markov process Y on H_α. Then, by a result of Wittig [56] under a certain relation of the covariance of Y to the covariance of a Gaussian martingale M there is an evolution operator $V_{-\alpha}(t,s)$ on $H_{-\alpha}$ s.t.

$$Y(\cdot) = \int_0^{\cdot} V_{-\alpha}(\cdot,s)dM(s) .$$

2. A more general class of processes with independent increments on a nuclear space has been investigated by Ustunel [51].

In the following sections we shall restrict ourselves to real separable Hilbert spaces H_α .

3.1 Examples of Stochastic Parabolic Equations on a Bounded Domain

Let D be a bounded open subset of R^n, $n \in N$, whose boundary ∂D is a C^∞-hypersurface, which lies on one side of D . Let

$$\mathring{A} = \sum_{i,j=1}^{d} \partial_i a_{ij} \partial_j + c \qquad (3.1.1)$$

be a strongly elliptic operator on

$$H_o := L_2(D)$$

with $a_{ij}(=a_{ji})$, $c \in C^\infty(\bar{D})$ (infinitely often differentiable on $\bar{D}$, real valued). We close $\mathring{A}$ w.r.t. homogeneous Dirichlet boundary conditions and denote this closure of $\mathring{A}$ by A . Then $-A$ is coercive (cf. Lions and Magenes [38] which can be verified by partial integration of $<-A\varphi,\psi>_o$, $\varphi\ \psi \in \mathrm{Dom}(A)$. Hence,

$A{\sim}U(t))) \in G(1,\beta)$, and $U(t)$ is analytic on H_o, where

$$\beta > \sup_{r\in D}|c(r)|$$

(cf. Tanabe [47], Th. 3.6.1). Moreover, A and $U(t)$ are self-adjoint.
Set

$$C_A^\infty(\bar{D}) := \{\varphi \in C^\infty(\bar{D}) : A^m\varphi|_{\partial D} \equiv 0 \quad m \in N\cup\{0\}\}$$

and

$$(H_m, <\cdot,\cdot>_m) := \text{closure of } C_A^\infty(\bar{D}) \text{ w.r.t. } <\cdot,\cdot>_m$$

where

$$<\varphi,\psi>_m := <(\beta-A)^m\varphi,\psi>_o = <(\beta-A)^{\frac{m}{2}}\varphi,(\beta-A)^{\frac{m}{2}}\psi>_o$$

$$\varphi,\psi \in C_A^\infty(\bar{D}) .$$

$<\cdot,\cdot>_m$ is a scalar product ($\beta-A$ is invertible).

For $\alpha \in R_+$ we define H_α by

$$H_\alpha := [H_{m+1}, H_m]_{1-\alpha} \quad \alpha \in [m,m+1],\ m \in N\cup\{0\} .$$

which is the intermediate space of Lions and Magenes [38].

$H_\alpha = \mathrm{Dom}((\beta-A)^{\frac{\alpha}{2}})$, $\alpha \in R_+$, is a real separable Hilbert space with scalar product

$$<\varphi,\psi>_\alpha := <(\beta-A)^{\frac{\alpha}{2}}\varphi,\ (\beta-A)^{\frac{\alpha}{2}}\psi>_o .$$

Now let us recall the definitions of the (standard) Sobolev spaces

$$H^m := (H^m(D);<<\cdot,\cdot>>_m) \quad m \in N\cup\{0\} ,$$

which are the spaces of real valued functions, m times differentiable in the generalized sense with all derivatives up to order m square integrable. $<<\cdot,\cdot>>_m$, the standard scalar product on H^m, is defined by

$$<<\varphi,\psi>>_m := \sum_{o\le|j|\le m} \int_D \partial^j\varphi\partial^j\psi ,$$

with $\partial^j = \partial_{j_1}\ldots\partial_{j_n}$, $j=(j_1\ldots j_n)$, $|j| = j_1+\ldots+j_n$

(we shall not explicitly write the integration variable and the integrator if there is no risk of confusion).

For $m=0$ $(H^O,<<\cdot,\cdot>>_o) = (H_o,<\cdot,\cdot>_o)$. Set

$$H^\alpha := [H^{m+1}, H^m]_{1-\alpha},\ \alpha \in [m,m+1],\ m \in N \cup \{0\}.$$

<u>Lemma 3.1.1</u>
For all $\alpha \in R_+$, H_α is a closed subspace of H^α with $<<\cdot,\cdot>>_\alpha$ restricted to H_α being equivalent to $<\cdot,\cdot>_\alpha$.

The <u>proof</u> follows for $\alpha \in N$ from Ladyženskaja [37] and for $\alpha \in R_+ \setminus N$ by interpolation (cf. Kotelene [34]).

□

Corollary 3.1.1

(I) The imbedding $H_\alpha \to H_\beta$ is Hilbert Schmidt, $\alpha > \beta + \frac{n}{2}$, $\beta \geq 0$. (i.e. $(i_\beta^\alpha; H_\alpha, H_\beta)$ is an abstract Wiener space)

(II) $H_\alpha \subset C^m(\overline{D})$ whenever $\alpha > m + \frac{n}{2}$, $m \in N \cup \{0\}$.

(III) For arbitrary $m \in N$ there exists an operator P_m s.t.

$P_m \in L(H_\alpha, H^\alpha(R^n))$, $\alpha \leq m$

$P_m \varphi = \varphi$ Lebesque a.e. on D $\varphi \subset H_0$

The proof follows from the corresponding statements for H^α (cf. Triebel [48], Lions and Magenes [38]).

□

We set

$$\Phi := \bigcap_{\alpha \geq 0} H_\alpha$$

and endow Φ with the locally convex topology defined by $\langle \cdot, \cdot \rangle_\alpha$, $\alpha \geq 0$. By Cor. 3.1.1, (I), Φ becomes a nuclear Fréchet space with this topology (cf. Schaefer [46]).

Now define

Φ' to be the strong dual of Φ

and

$$H_{-\alpha} := \{\varphi' \in \Phi' \ |\varphi'|_\alpha := \sup_{|\psi|_\alpha \leq 1, \psi \in \Phi} |\varphi'(\psi)| < \infty \}.$$

Remark 3.1.1

(I) Obviously, alle elements φ' from $H_{-\alpha}$ can be extended into continuous linear functions on H_α whence $H_{-\alpha}$ becomes the (distributional) strong dual of H_α. Moreover, we may identify H_0 with its strong dual H_0' which implies the chain of (dense) continuous inclusions

$$\Phi \subset H_\alpha \subset H_\gamma \subset H_0 = H_0' \subset H_{-\gamma} \subset H_{-\alpha} \subset \Phi' \quad (3.1.2)$$

for $\alpha \geq \gamma \geq 0$ (cf. Gel'fand and Vilenkin [20]).

(II) If $\varphi, \psi \in H_0$, then it is easy to see that

$$\langle\varphi,\psi\rangle_{-\alpha} = \langle(\beta-A)^{\frac{-\alpha}{2}}\varphi,\ (\beta-A)^{\frac{-\alpha}{2}}\psi\rangle_o$$

(III) From (I) and Cor. 3.1.1, (I), we obtain

$(i^\gamma_\alpha, H_\gamma, H_\alpha)$, ($i^\gamma_\alpha$ the imbedding of H_γ into H_α)

is an abstract Wiener space, whenever $\alpha + \frac{n}{2} < \gamma, \alpha, \gamma \in \mathbb{R}$ (cf. Kuo [36]). Indeed, by Cor. 3.1.1 there is a CONS of eigenvectors $\{\varphi_l\} \subset C^\infty_A(\overline{D})$ for $(\beta-A)$ with eigenvalues $\lambda_\ell > O$ Hence, $\varphi^\gamma_l := \varphi_l\lambda_l^{-\gamma/2}$ is a CONS for H_γ, $\gamma \in R$, implying $\sum_\ell |\varphi^\gamma_\ell|^2_\alpha = \sum_\ell |\varphi^{\gamma-\alpha}_\ell|^2_o < \infty$ iff $\gamma > \frac{n}{2} + \alpha$. Moreover, we have

$$H_\gamma = \{\varphi'\in\Phi' : \Sigma[\varphi'(\varphi_l)]^2\lambda^\gamma_l < \infty\}.$$

(IV) DaPrato and Grisvard [13] have given a concrete representation of $H_{-\alpha}$ as the quotient space $H^2_o/G((\beta-A)^{\frac{\alpha}{2}})$, where $G((\beta-A)^{\frac{\alpha}{2}})$ is the graph of $(\beta-A)^{\frac{\alpha}{2}}$.

The following lemma shows that U(t) and A have nice extensions (resp. restrictions) to H_α, $\alpha \in R$.

<u>Lemma 3.1.2</u>

For any $\alpha < 0$) U(t) and A have extensions (resp. restrictions) $U_\alpha(t)$ and A_α on H_α such that:

$A_\alpha \sim U_\alpha(t) \in G(1,\beta)$, $Dom(A_\alpha) = H_{\alpha+2}$, and $U_\alpha(t)$ is analytic.

The proof is given in Kotelenez [34].

□

Let M be an $H_{-\alpha}$-valued l.s.i. cadlag martingale $(\alpha \geq 0)$ and consider the stochastic evolution equation:

$$\left.\begin{aligned} dY(t) &= A_{-\alpha}Y(t)dt + dM(t) \\ Y(o) &\in H_{-\alpha} \end{aligned}\right\} \quad (3.13)$$

As a consequence of Lemma 3.1.2 and Th. 2.1.4 (resp. Th. 2.1.3) we obtain:

Corollary 3.1.2

(3.1.3) has a (unique) mild solution

$$Y(\cdot) = U_{-\alpha}(\cdot)Y_o + \int_0^{\cdot} U_{-\alpha}(\cdot-s)dM(s)$$

s.t.

$$Y \in \begin{cases} L_2^F([0,T]x\Omega;H_{-\alpha+1}) & \text{if } M \text{ square integrable,} \\ D([0,\infty);H_{-\alpha}) \text{ a.s.} & \text{it } M \text{ is cadlag} \\ C^{\mu}([0,T];H_{-\alpha}) \cap C((0,T];H_{-\alpha+\varepsilon}) \text{ a.s.} & \\ \quad \text{for all } T>0,\ \mu \in [0,\frac{1}{2}),\ \varepsilon \in [0,1) & \\ \quad\quad \text{if } M \text{ satisfies A.2.1.3)} & \end{cases} \quad (3.1.4)$$

Example 3.1.1

Let $a_{ij} = 0$ if $i \neq j$ in (3.1.1) $c = b - d$, where $b,d \in C^{\infty}(\overline{D},R_+)$ and $c_o \in C^{\infty}(\overline{D},R_+)$. Moreover, let $X_o \in C^{3+\mu}(\overline{D},R_+)$ $(\mu \in (0,1)$ and assume

$$X_o|_{\partial D} \equiv 0\ , \quad \mathring{A}X_o + c_o|_{\partial D} \equiv 0\ , \quad \mathring{A}^2X_o + c_o|_{\partial D} \equiv 0\ .$$

Under these assumptions there is a unique positive solution X to the deterministic PDE

$$\left.\begin{aligned} \frac{\partial}{\partial t}X(t,r) &= AX(t,r) + c_o(r) \\ X(o) &= X_o \end{aligned}\right\} \quad (3.1.5)$$

which can be obtained as the high density limit (or thermodynamic limit) of a system branching diffusions (with birth rate $b(r)$ death rate $d(r)$, immigration $c_0(r)$ and emigration (or leaking) due to the boundary conditions) (Kotelenez [34]).

Set

$$F_L(X(t)) := \sum_{i=1}^{n} -2\partial_i X(t)\alpha_i\partial_i$$

$$F_R(X(t)) := (b+d)X(t) + c_o$$

and let M_J be the (independent) Gaussian martingales with characteristic functional

$$\exp(\frac{1}{2}\int_0^t \langle F_J(x(s)\varphi,\varphi\rangle_0 ds), \ \varphi \in \Phi \ , \ J \in \{L,R\}.$$

Then, by Itô [27], M_L lives on $H_{-\alpha}$ and M_R on $H_{-\alpha+1}$ for all $\alpha > \frac{n}{2} + 1$. Setting

$$M = M_L + M_R$$

in (3.13) we obtain the fluctuation process Y for the branching diffusions around X as the mild solution to (3.1.3). Moreover, in this case

$$Y \in C^{\mu}([0,T];H_{-\alpha}) \cap C((0,T];H_{-\alpha+1}) \quad \text{a.s.} \qquad (3.1.6)$$

for all $\alpha > \frac{n}{2} + 1$ (cf. Kotelenez [34]).

Example 3.1.2
A similar result was obtained in Kotelenez [31] for $D = [0,1]$ with Neumann boundary conditions.

Example 3.1.3
Let $D = [0,L]$, $L > 0$, $\mathring{A} = \frac{\partial^2}{\partial x^2} - 1$, and let M be cylindrical Brownian motion on $L_2(0,L)$. Clearly, M is a Gaussian martingale on $H_{-\alpha}$ for all $\alpha > \frac{1}{2}$. The mild solution Y to (3.1.3) was obtained by Walsh [52] as the diffusion approximation to a stochastic neural response model. In particular, (3.1.4) shows that

$$Y \in C((0,T];H_{\varepsilon}) \quad \text{a.s.} \quad \text{for all} \quad \varepsilon < \frac{1}{2} ,$$

i.e. Y is function valued. However, by using multi parameter estimates Walsh (loc. cit) has shown that as a function of $(t,r) \in [0,T] \times [0,L]$ has a modulus of continuity of the order of $\delta^{\frac{1}{4}}(\log\delta^{-1})^{\frac{1}{2}}$ with $\delta = \{(t-s)^2 + (r-q)^2\}^{\frac{1}{2}}$. In particular, $Y(t,r)$ has Hölder modulus $\frac{1}{4}$ on $[0,T] \times [0,L]$ for all $T > 0$.

Remark 3.1.1

We can completely analogously construct the nuclear space (3.1.2) associated with an arbitrary self-adjoint strongly elliptic operator $\tilde{A}$ of order 2m, m ≥ 1, and show the extendibility (resp. restrictibility) of $\tilde{A}$ and $\tilde{U}(t)$ (the semigroup generated by $\tilde{A}$) to $\tilde{A}_\gamma$ and $\tilde{U}_\gamma(t)$ onto $\tilde{H}_\gamma$ where $\tilde{H}_\alpha = \mathrm{Dom}(\beta-A)^{\frac{\alpha}{2m}})$. Let M be an $H_{-\alpha}$-valued martingale satisfying A.2.1.3). Then Th. 2.1.3 yields that

$$\left.\begin{array}{l} d\tilde{Y}(t) = \tilde{A}_{-\alpha}\tilde{Y}(t)dt + dM(t) \\ \tilde{Y}(0) \in H_{-\alpha} \end{array}\right\} \quad (3.1.7)$$

has a unique mild solution $Y(\cdot) = \tilde{U}_{-\alpha}(\cdot)\tilde{Y}_0 + \int_0^{\cdot} \tilde{U}_{-\alpha}(\cdot-s)dM(s) \in C^\mu([0,T];H_{-\alpha}) \cap C((0,T];\tilde{H}_{-\alpha+\varepsilon})$

for all $\mu \in [0,\frac{1}{2})$, $\varepsilon \in [0,m)$

(cf. Agmon [1]). Thus, if $2m > n$ and M lives on $\tilde{H}_{-\alpha}$ for all $\alpha > \frac{n}{2}$, then

$$\tilde{Y} \in C([0,T];\tilde{H}_\varepsilon) \quad \text{a.s. for all } \varepsilon \in [0,m-\tfrac{n}{2})$$

i.e. $\tilde{Y}$ is function valued. In particular, M is a cylindrical Brownian motion B on $\tilde{H}_0$ (since $(i^0_{-\alpha},\tilde{H}_0,\tilde{H}_{-\alpha})$ is an abstract Wiener space for $\alpha > \frac{n}{2}$) . In this case one can also consider the nonlinear stochastic evolution equation

$$\left.\begin{array}{l} d\tilde{Y}(t) = \tilde{A}_{-\alpha}\tilde{Y}(t)dt + B(t,\tilde{y}))dt + C(t,\tilde{y}(t))dB(t) \\ \tilde{Y}(o) \in H_0 \end{array}\right\} \quad (3.1.8)$$

for certain B and C . This was done by Dawson [17], [18], and Funaki [19]. Let us look only at Funaki's results and assume

A.3.1.1)

For arbitrary $T > 0$

(i) $C(\cdot,\cdot) : [0,T] \times H_0 \longrightarrow L(H_0)$ s.t.

(i.1) $C^*(\cdot,\cdot)\varphi_\ell \in C([0,T]\times H_0;H_0)$ $\forall \ell$, where $\{\varphi_\ell\}$ are the eigenvectors of $\tilde{A}$ and C^* is the adjoint of $C(t,x)$.

Moreover, there is a $K < \infty$ s.t.

(i.2) $|C(t,x)|_{L(H_0)} \leq K$

(i.3) $|(C^*(t,x)-C^*(t,y))\varphi_\ell|_0 \leq K|x-y|_0$ $\forall$ $x,y \in H_0$, $\ell \in N$

(ii) $B(\cdot,\cdot) \in C([0,T]\times H_0;H_0)$ and

(ii.2) $|B(t,x) - B(t,y)|_0 \leq K|x-y|_0$ $\forall x,y \in H_0$.

(ii.3) $|B(t,x|_0 \leq K$ $\forall x \in H_0$

Under A.3.1.1) Funaki [19] has shown that (3.1.8) has a unique solution Y which is jointly continuou if $\tilde{Y}_0(r)$ is continuous. The key to this result is the estimate for $X(t) := \tilde{Y}(t) - \tilde{U}(t)\tilde{Y}_0$:

For any (large) $p > 0$ there is a $C < \infty$ s.t.

$$E(|X(t,r)-X(s,q)|^{2p}) \leq C\{|r-q|^\gamma+|t-s|^{1-\frac{n}{2m}}\}^p \qquad (3.1.9)$$

with $0 < \gamma < 2m-n$.

However, using Ibragimov's extension of Kolmogorov's theorem to the multiparameter case (cf. [25], Th. 6 (3.1.9) implies the even better

Theorem 3.1.1

Assume A.3.1.1) and that $\tilde{Y}(o,r)$ has Hölder modules $\frac{1}{4}$ on $\bar{D}$. Then, for any $T > 0$

$\tilde{Y}(t,r)$ is jointly Hölder continuous on $[0,T]\times\bar{D}$ for all Hölder exponents $\alpha < \frac{2m-n}{4m}$.

Remark 3.1.2

Th. 3.1.1 implies in particular for $m = n = 1$ that the solution of the nonlinear SPDE (3.1.8) has the same Hölder modulus on $[0,T]\times\bar{D}$ as the solution of Walsh's linear SPDE (cf. Ex. 3.1.3).

3.2 Examples of Stochastic Parabolic Equations on R^n

The standard nuclear Gel'fand triple for this case is the Schwartz triple

$$S(R^n) \subset L_2(R^n) \triangleq L_2'(R^n) \subset S'(R^n) \qquad (3.2.1)$$

A description of (3.2.1) with "intermediate" Hilbert spaces H_α can be found in Holley and Stroock [24]. Let us briefly recall how the H_α are generated. Set

$$f_k(x) := (-1)^k e^{\frac{x^2}{2}} \partial_x^k e^{-x^2}, \quad x \in R$$

and

$$g_k(x) := (\Pi^{\frac{1}{2}} 2^k k!)^{-\frac{1}{2}} f_k(x) \quad x \in R$$

g_k is called the k-th Hermite function and $\{g_k\}$ are a CONS in $L_2(R)$. Set

$$h_k(x) := g_{k_1}(x_1)\cdot\ldots\cdot g_{k_n}(x_n), \quad x \in R^n \qquad (*)$$

$$x = (x_1,\ldots,x_n) \in R^n, \ k = (k,\ldots,k_n) \quad (N\cup\{0\})^n.$$

Then, $\{h_k\}$ is a CONS in $H_0 := L_2(R^n)$ and all $h_k \in S := S(R^n)$. Moreover, if $|x|^2$ denotes the multiplication operator ($x \in R^n$, $|\cdot|$ Euclidean norm) and Δ the Laplacian on R^n then

$$(-\Delta+|x|^2)h_k = (2|k|+n)h_k, \qquad (3.2.2)$$

where $|k| = k_1+\ldots+k_n$. Thus $(-\Delta+|x|^2)$ plays the role of $(\beta-A)$ of Section 3.1 and (omitting the R^n in $S(R^n)$ etc.) we obtain

$$S \subset H_\alpha \subset H_\gamma \subset H_0 = H_0' \subset H_{-\gamma} \subset H_{-\alpha} \subset S' \qquad (3.2.3)$$

for $\alpha \geq \gamma \geq 0$ where

$$H_\alpha := \{\varphi \in S' : |\varphi|_\alpha := |(-\Delta+|x|^2)^{\frac{\alpha}{2}}\varphi|_0 < \infty\}, \ \alpha \in R$$

From (3.2.2) we obtain

Remark 3.2.1
The imbedding $H_\alpha \longrightarrow H_\beta$ is Hilbert Schmidt, wheneve $\alpha > \beta + n$ (i.e. $(i^\alpha_\beta, H_\alpha, H_\beta)$ is an abstract Wiener space).

In view of the following applications we would like to solve the analogue of (3.1.3) with $\frac{1}{2}\Delta$ instead of A. Therefore, denoting by $S(t)$ the semigroup generated by $\frac{1}{2}\Delta$ we show:

Lemma 3.2.1
$S(t)$ is $G(1,\beta)$ extendible (restrictible) onto H_α for all $\alpha \in R$.

Proof
(i) Let $\alpha \in R$. Then we easily see

$$h^\alpha_k := (2|k|+n)^{-\frac{\alpha}{2}} h_k \quad \text{is a CONS in } H_\alpha .$$

Thus, for $\varphi, \psi \in S$

$$<\varphi,\psi>_\alpha = \sum_k <\varphi,h^\alpha_k>_\alpha <\psi,h^\alpha_k>_\alpha$$

$$= \sum_k <\varphi,h_k>_0 <\psi,h_k>_0 \ (2|k|+n)^\alpha$$

Hence, for $\varphi, \psi \in S$

$$<\Delta\varphi,\psi>_\alpha = \sum_k <\varphi,\Delta h_k>_0 <\psi,\Delta h_k>_0 (2|k|+n)^\alpha ,$$

and all we have to show is the existence of a $\beta_\alpha < \infty$ s.t.

$$<\Delta\varphi,\varphi>_\alpha \le \beta_\alpha |\varphi|^2_\alpha \quad \text{for} \quad \varphi \in S$$

which is equivalent to

$$|S(t)\varphi|^2_\alpha \le e^{\beta_\alpha t} |\varphi|^2_\alpha , \quad \varphi \in S .$$

This last relation implies the (unique) extendibility (resp. restrictibility) of $S(t)$ to $S_\alpha(t)$ onto H_α with generator $\frac{1}{2}\Delta_\alpha$ (the extension, resp. restriction, of $\frac{1}{2}\Delta$).

(ii) To simplify notation assume $n = 1$ and note that

$$\partial_x^2 g_k(x) = \frac{(k(k-1))^{\frac{1}{2}}}{2} g_{k-2}(x) - \frac{2k+1}{2} g_k(x) + \frac{[(k+1)(k+2)]^{\frac{1}{2}}}{2} g_{k+2}(x)$$

$$x \in R, \quad k \in N \cup \{0\}, \; g_{-1} \equiv g_{-2} \equiv 0$$

(cf. Holley and Stroock [24], (A8), (A.9)).

Thus, for $\varphi \in S(R)$, setting $\varphi_m^\alpha := \langle \varphi, g_m^\alpha \rangle_\alpha$, we have $\langle \Delta\varphi, \varphi \rangle_\alpha$

$$= \sum_m \{ \varphi_m^\alpha \varphi_{m+2}^\alpha (2m+1)^{\frac{\alpha}{2}} (2m+5)^{-\frac{\alpha}{2}} \frac{[(m+1)(m+2)]^{\frac{1}{2}}}{2} - (\varphi_m^\alpha)^2 \left(\frac{2m+1}{2}\right) + \varphi_m^\alpha \varphi_{m-2}^\alpha (2m+1)^{\frac{\alpha}{2}} (2m-3)^{-\frac{\alpha}{2}} \frac{[[m-1)m]^{\frac{1}{2}}}{2} \}$$

(where we put all terms equal to 0 if the index is smaller than 0)

$$\leq \frac{1}{2} \sum_m (\varphi_m^\alpha)^2 \cdot \gamma_m^\alpha$$

with $$\gamma_m^\alpha := -(2m+1) + \frac{[m(m-1)]^{\frac{1}{2}}}{2} \left[\left(\frac{2m-3}{2m+1}\right)^{\frac{\alpha}{2}} + \left(\frac{2m+1}{2m-3}\right)^{\frac{\alpha}{2}} \right]$$

$$\frac{[(m+1)(m+2)]^{\frac{1}{2}}}{2} \left[\left(\frac{2m+1}{2m+5}\right)^{\frac{\alpha}{2}} + \left(\frac{2m+5}{2m+1}\right)^{\frac{\alpha}{2}} \right]$$

and since it is easy to see that

$$\sup_m \gamma_m^\alpha < \infty$$

we may take

$$\beta_\alpha := \frac{1}{2} \sup_m \gamma_m^\alpha .$$

(iii) The n-dimensional case is easily reduced to the onedimensional case by (*) and (3.2.2)

□

Now we can solve the analogue of (3.1.3). Let M be an $\mathcal{H}_{-\alpha}$-valued ℓ.s.i. cadlag martingale ($\alpha \geq 0$) and consider for $a \in R$

$$\begin{aligned} dY(t) &= (\Delta_{-\alpha}+a)y(t)dt + dM(t) \\ Y(o) &\in \mathcal{H}_{-\alpha} \end{aligned} \tag{3.2.4}$$

As a consequence of Lemma 3.2.1 and Th. 2.1.1 (or Remark 2.1.1) we obtain (setting $T_{-\alpha}(t) := S_{-\alpha}(t)e^{at}$):

<u>Corollary 3.2.1</u>
(3.2.4) has a (unique) mild solution Y s.t.

$$Y(\cdot)=T_{-\alpha}(\cdot)Y(0)+\int_O^\cdot T_{-\alpha}(\cdot-u)dM(u)$$

$$\in \begin{cases} D([0,\infty);\mathcal{H}_{-\alpha}) \text{ a.s. if } M \text{ is cadlag} \\ C([0,\infty);\mathcal{H}_{-\alpha}) \text{ a.s. if } M \text{ is continous} \end{cases} \tag{3.2.5}$$

<u>Remark 3.2.2</u>
Since $\mathcal{H}_\alpha$ are not generated by Δ alone (but by $(-\Delta+|x|^2)$ -unlike the H_α in Section 3.1) we cannot as in Section 3.1 obtain spatial regularity for the mild solution of (3.2.4) in terms of $\mathcal{H}_\alpha$.

<u>Example 3.2.1</u>
Let $a \in R$ and $m_2, V > 0$. Let M be the $\mathcal{S}'$-valued Gaussian martingale with covariance $\int_O^{t\wedge s} \langle F(u)\varphi,\psi\rangle_O du$ where $F(u) = -(\nabla e^{at}\nabla)+(m_2V-a)e^{at}$.

From Itô [27] and Remark 3.2.1 we obtain that M

lives as a continuous process on $\mathcal{H}_{-\alpha}$ for all $\alpha > n + 1$. Hence the mild solution Y to (3.2.4) lives as a continous Gaussian process on $\mathcal{H}_{-\alpha}$ for all $\alpha > n + 1$ (Cor. 3.2.1). This process was obtained by Gorostiza [22] as the fluctuation limit for an infinite system of unscaled branching Brownian motions. V is the parameter of the exponential lifetime distribution, m_1 is the mean of the branching law, m_2 the second factorial moment and $a := V(m_1-1)$.

Example 3.2.2

If M is the cylindrial Brownian motion on $\mathcal{H}_0$ (i.e., its covariance is $t \wedge s \langle \varphi, \psi \rangle_0$, $\varphi, \psi \in \mathcal{S}$) then from Itô and Remark 3.2.1 M is an $\mathcal{H}_{-\alpha}$-valued Gaussian martingale for all $\alpha > n$. Then for $a = o$ the mild solution to (3.2.4) describes the fluctuation limit obtained by Holley and Stroock [24] for an infinite system of critial branching Brownian motions by scaling space and time.

Remark 3.2.3

(i) Another example of (3.2.4) can be found in Itô [28].

(ii) Walsh [53] has investigated various scaling limits for systems of branching Brownian motions on $\mathcal{S}'$ and shown that the martingale part can be decomposed into two orthogonal martingales which account for diffusion and branching, respectively (cf. also Kotelenez [34] for branching diffusions on a lattice).

Remark 3.2.4

Holley and Stroock [24] have also investigated the Ornstein-Uhlenbeck process with M(t) cylindrical Brownian motion and Δ substituted by $\Delta - |x|^2$ in (3.2.4). Clearly, in this case the extension (restriction) procedure becomes much easier - it is exactly the same as in Section 3.1. In particular, the analogue of Lemma 3.1.2 holds, with H_α substituted by $\mathcal{H}_\alpha$.

3.3 Examples of Second Order Equations on a Bounded Domain

Let $D \subset \mathbb{R}^n$ and A be as in Section 3.1. Moreover assume

A.3.3.1)
A is negative definite.

Under assumption A.3.3.1) we may take $\beta = 0$ in the construction of the H_α spaces.

Let us consider the unperturbed second order equation:

$$\frac{\partial^2}{\partial t^2} Y(t,r) = AY(t,r) - \mu \frac{\partial}{\partial t} Y(t,r) \ , \quad \mu \geq 0. \quad (3.3.1)$$

Analogously to the parabolic case we make an extension procedure for the associated first order equation

$$\frac{\partial}{\partial t} X(t,r) = \hat{A} X(t,r)$$

$$X(t) := \begin{pmatrix} Y(t) \\ \frac{\partial}{\partial t} Y(t) \end{pmatrix} , \ \hat{A} := \begin{pmatrix} 0 & I \\ A & -\mu \end{pmatrix} \quad (3.3.2)$$

Note that by Curtain and Pritchard [11]

$$\hat{A} \sim \hat{U}(t) \in G(1,0)$$

$$\text{on} \quad (\hat{H}_0, (\cdot,\cdot)_0) = (H_1 \times H_0, \langle\cdot,\cdot\rangle_1 + \langle\cdot,\cdot\rangle_0) \ .$$

Then, we define for $\alpha \in R$

$$(\hat{H}_\alpha, (\cdot,\cdot)_\alpha) := (H_{\alpha+1} \times H_\alpha, \langle\cdot,\cdot\rangle_{\alpha+1} + \langle\cdot,\cdot\rangle_\alpha)$$

and obtain as in Section 3.1 a nuclear Gel'fand triple for (3.3.1):

$$\Phi \subset \hat{H}_\alpha \subset \hat{H}_\gamma \subset \hat{H}_0 = \hat{H}_0' \subset \hat{H}_{-\gamma} \subset \hat{H}_{-\alpha} \subset \hat{\Phi}' \ . \quad (3.3.3)$$

Lemma 3.3.1
$\hat{U}(t)$ is $G(1,0)$ extendible (resp. restrictible) on $\hat{H}_\alpha$ for all $\alpha \in R$.

Proof

(i) One easily verifies $\hat{U}(t)\hat{\Phi} \subset \hat{\Phi}$

(ii) If $\hat{\varphi} \in \hat{\Phi}$ then

$$(\hat{U}(t)\hat{\varphi},\hat{U}(t)\hat{\varphi})_\alpha = (\hat{\varphi},\hat{\varphi})_\alpha + \int_0^t 2(\hat{A}\hat{U}(s)\hat{\varphi},\hat{U}(s)\hat{\varphi})_\alpha ds$$
$$\leq (\hat{\varphi},\hat{\varphi})_\alpha$$

since for arbitrary $(\varphi,\psi) \in \hat{H}_\alpha = H_{\alpha+1} \times H_\alpha$

$$(\hat{A}(\varphi,\psi),(\varphi,\psi))_\alpha = (\varphi,\psi)_{\alpha+1} + \langle A\varphi,\psi\rangle_\alpha - \beta|\psi|^2_\alpha$$
$$= -\langle A\varphi,\psi\rangle_\alpha + \langle A\varphi,\psi\rangle_\alpha - \beta|\psi|_\alpha = -\beta|\psi|^2_\alpha$$

by the construction of H_α (cf. Section 3.1)

(iii) From this the lemma follows as in Section 3.2. □

Now let M be an $H_{-\alpha}$-valued martingale and consider the (formal) equation

$$\left.\begin{aligned} &\frac{\partial^2}{\partial t^2}Y(t,r) = AY(t,r) - \mu\frac{\partial}{\partial t}Y(t,r) + \frac{\partial}{\partial t}M(t) \\ &Y(o,r) = Y_o(r), \frac{\partial}{\partial t}Y(o,r) = Y_1(r) \end{aligned}\right\} \quad (3.3.4)$$

which has a meaning as the evolution equation:

$$\left.\begin{aligned} &dX(t) = \hat{A}_{-\alpha}X(t)dt + \begin{pmatrix} 0 \\ dM(t) \end{pmatrix} \\ &X(o) = \begin{pmatrix} Y_o \\ Y_1 \end{pmatrix} . \end{aligned}\right\} \quad (3.3.5)$$

Corollary 3.3.1

(3.3.5) has a (unique) mild solution

$$\left.\begin{aligned} &X(\cdot) = \hat{U}_{-\alpha}(\cdot)X(o) + \int_o^{\cdot} \hat{U}_{-\alpha}(\cdot - s)d\begin{pmatrix} 0 \\ M(s) \end{pmatrix} \\ &\in \begin{cases} D([0,\infty);\hat{H}_{-\alpha}) \text{ a.s. if } M \text{ is cadlag} \\ C([0,\infty);\hat{H}_{-\alpha}) \text{ a.s. if } M \text{ is continous} \end{cases} \end{aligned}\right\} (3.3.6)$$

Example 3.3.1
Let $D = [0,L] \subset \mathbb{R}$, $A := \lambda\Delta$, $\lambda > 0$, $M(t) = \sigma W(t)\delta(x-a)$, $\sigma > 0$, $a \in [0,L]$, δ the Delta function, and $W(t)$ one-dimensional standard Brownian motion. This is an example from continuum mechanics where a string is exposed to a random excitation by a point force in a (cf. Wedig [54]). From Remark 3.1.1 (III) we obtain that $M(t)$ lives on $\hat{H}_{-\alpha}$ for all $\alpha > \frac{1}{2}$. Hence by Corollary 3.3.1 $X = \begin{pmatrix} Y \\ \frac{\partial}{\partial t} Y \end{pmatrix}$ lives as a continous process on $H_{-\alpha}$, $\alpha > \frac{1}{2}$ which implies that the solution Y of (3.3.4) lives on H_ε for all $\varepsilon < \frac{1}{2}$. Moreover, a criterion of Zabczyk [57] yields the existence of a unique invariant measure for (3.3.4) under the preceding assumptions.

Other examples such as those given by Wedig in [55] can be handled in the same way.

Peter Kotelenez
Forschungsschwerpunkt Dynamische Systeme
Universität Bremen
Bibliothekstraße
Postfach 330 440
2800 Bremen 33
West Germany

4. REFERENCES

[1] S. AGMON: "Lectures on Elliptic Boundary Value Problems". Princeton, N. J.: D. van Nostrand Co. Inc. (1965)

[2] L. ARNOLD, R. F. CURTAIN, P. KOTELENEZ: "Nonlinear Evolution Equation in Hilbert Space". Forschungsschwerpunkt Dynamische Systeme, Universität Bremen, Report Nr. 17 (1980)

[3] A. BAGCHI, V. BORKAR: "Parameter Identification in Infinite Dimensional Linear Systems", Stochastics 12 (3+4) 201-213 (1984)

[4] A. V. BALAKRISHNAN: "On Abstract Stochastic Bilinear Equations with white Noise Inputs", Preprint (1983)

[5] H. T. BANKS, F. KAPPEL: "Spline Approximations for Functional Differential Equations", J. diff. Equations 34 496-522 (1979)

[6] P. BILLINGSLEY: "Convergence of Probability Measures", John Wiley & Sons, New York 1968

[7] P. L. BUTZER, H. BERENS: "Semi-Groups of Operators and Approximation" Springer-Verlag, Berlin-New York 1967

[8] P. R. CHERNOFF: "Two Counterexamples in Semi-group Theory on Hilbert Space", Proc. Amer. Math. Soc. 56 253-255 (1976)

[9] A. CHOJNOWSKA-MICHALIK: "Stochastic Differential Equation in Hilbert Spaces and their Applications", Ph. D. Thesis, Institute of Mathematics, Polish Academy of Science, 1976

[10] R. F. CURTAIN: "Markov Processes Generated by Linear Stochastic Evolution Equations", Stochastic 5, 135-165 (1981)

[11] R. F. CURTAIN, A. J. PRITCHARD: "Infinite Dimensional Linear Systems Theory", LN in Control and Information Sciences 8, Springer-Verlag, Berlin-New York 1978

[12] G. DAPRATO: "Regularity Results of a Convolution Stochastic Integral and Applications to Parabolic Stochastic Equations in a Hilbert Space", Scuola Normale Superiore, Pisa, 1982

[13] G. DAPRATO, P. GRISVARD: "Maximal Regularity for Evolution Equations by Interpolation and Extrapolation" to appear in J. functional Analysis

[14] G. DAPRATO, M. IANNELLI, L. TUBARO: "Some Results on Linear Stochastic Differential Equations in Hilbert Spaces" Stochastics 6, 105-116 (1982)

[15] G. DAPRATO, M. IANNELLI, L. TUBARO: "On the Path Regularity of a Stochastic Process in a Hilbert Space Defined by the Itô Integral", Stochastics 6, 315-322 (1982)

[16] E. B. DAVIES: "One-Parameter Semigroups", Academic Press, London-New York 1980

[17] D. A. DAWSON: "Stochastic Evolution Equations", Math. Biosciences 15, 287-316 (1972)

[18] D. A. DAWSON: "Stochastic Evolution Equations and Related Measure Processes", J. multivariate Analysis 5, 1-52 (1975)

[19] T. FUNAKI: "Random Motion of Strings and Related Stochastic Evolution Equations", Nagoya math. J. 89, 129-193 (1983)

[20] I. M. GEL'FAND, N. YA. VILENKIN: "Generalized Functions", Vol. 4, Academic Press, New York-London 1964

[21] I. I. GIHMAN, A. V. SKOROHOD: "The Theory of Stochastic Processes", Vol. 1, Springer-Verlag, Berlin-New York 1974

[22] L. GOROSTIZA: "High Density Limit Theorems for Infinite Systems of Unscaled Branching Brownian Motions", Ann. of Probab. 11 (2), 374-392 (1983)

[23] I. GYÖNGY: "On Stochastic Equations with Respect to Semimartingales III", Stochastics 7, 231-254 (1982)

[24] R. HOLLEY, D. W. STROOCK: "Generalized Ornstein-Uhlenbeck Processes and Infinite Particle Branching Brownian Motions", Publ. RIMS, Kyoto Univ. 14, 741-788 (1978)

[25] I. A. IBRAGIMOV: "On Smoothness Conditions for Trajectories of Random Functions", Theory Probab. Appl., Vol. 28 (2), 240-262 (1983)

[26] A. ICHIKAWA: "Semilinear Stochastic Evolution Equations: Boundedness, Stability and Invariant Measures", Stochastics 12, 1-39 (1984)

[27] K. ITÔ: "Continuous Additive *S*' Processes", in B. Grigelionis (ed.) "Stochastic Differential Systems" Springer-Verlag, Berlin-New York 1980

[28] K ITÔ: "Distribution-Valued Processes Arising from Independent Brownian Motions", Math. Z. 182, 17-33 (1983)

[29] J. JACOD: "Calcul Stochastique et Probléme de Martingales", LN in Mathematics 714, Springer-Verlag, Berlin-New York 1979

[30] P. KOTELENEZ: "A Submartingale Type Inequality with Applications to Stochastic Evolution-Equations", Stochastics 8, 139-151 (1982)

[31] P. KOTELENEZ: "Law of Large Numbers and Central Limit Theorem for Chemical Reactions with Diffusions" Ph. D. Thesis, Bremen 1982

[32] P. KOTELENEZ: "Continuity Properties of Hilbert Space Valued Martingales", Stochastic Processes Appl. 17 (1), 115-125 (1984)

[33] P. KOTELENEZ: "A Stopped Doob Inequality for Stochastic Convolution Integrals and Stochastic Evolution Equations" Stoch. Analysis and Applications 2 (3), 245-265 (1984)

[34] P. KOTELENEZ: "Linear Parabolic Differential Equations as Limits of Space-Time Jump Markov Processes" Technical Report Series of the Laboratory for Research in Statistics and Probability No 38, Carleton University (1984)

[35] N. V. KRYLOV, B. L. ROZOVSKIJ: "On Stochastic Evolution Equations", Itogi Nauki i Tehniki, Ser. Sov. Probl. Mat. 14 71-146 (1979) (in Russian)

[36] H.-H. KUO: "Gaussian Measures in Banach Spaces" Springer-Verlag, Berlin-New York 1975

[37] O. A. LADYŽENSKAYA: "The Mixed Problem for a Hyperbolic Equation", GITTL, Moscow 1953 (in Russian)

[38] J. L. LIONS, E. MAGENES: "Non-Homogeneous Boundary Value Problem I", Springer-Verlag, Berlin-New York 1972

[39] M. METIVIER: "Semimartingales", Walter de Gruyter, Berlin-New York 1982

[40] M. METIVIER, J. PELLAUMAIL: "On a Stopped Doob's Inequality and General Stochastic Equations", Ann. of Probab. 8, Nr. 1, 96-114 (1980)

[41] M. METIVIER, J. PELLAUMAIL: "Stochastic Integration", Academic Press, New York 1980

[42] S.-I. NAKAGIRI: "On the Fundamental Solution of Delay-Differential Equations in Banach Spaces", J. diff. Equations 41, 349-368 (1981)

[43] E. PARDOUX: "Equations aux derivees partielles stochastic non linéaires monotones. Etude de solutions fortes de type Itô", These doct. Sci. math. Univ. Paris Sud, 1975

[44] E. PARDOUX: "Stochastic Partial Differential Equations and Filtering of Diffusion Processes", Stochastics 3, 127-167 (1979)

[45] B. L. ROZOVSKIJ: "Stochastic Evolution Systems", Nauka, Moscow 1983 (in Russian)

[46] H. H. SCHAEFER: "Topological Vector Spaces", Springer-Verlag, New York-Berlin 1980

[47] H. TANABE: "Equations of Evolution", Pitman, London-San-Francisco-Melbourne 1979

[48] H. TRIEBEL: "Interpolation Theory, Function Spaces, Differential Operators", North-Holland, Amsterdam-New York-Oxford 1978

[49] L. TUBARO: "An Estimate of Burkholder Type for Stochastic Processes Defined by the Stochastic Integral", Stoch. Analysis and Applications, 187-192 (1984)

[50] C. TUDOR: "Some Properties of Mild Solutions of Delay Stochastic Evolution Equations", Preprint (1984)

[51] A. S. USTUNEL: "Additive Processes on Nuclear Spaces" Ann. of Probab. 12, Nr. 3, 858-868 (1984)

[52] J. WALSH: "A Stochastic Model of Neural Response" Adv. Appl. Prob. 13, 231-281 (1981)

[53] J. WALSH: Preprint (1984)

[54] W. WEDIG: "Covariance Analysis of Distributed Systems under Stochastic Point Forces" Preprint 1982

[55] W. WEDIG: "Moments and Probability Densities of Parametrically Excited Systems and Continuous Systems" Abhandlungen der Adademic der Wissenschaften der DDR Abt. Mathematik-Naturwissenschaften-Technik 6N (1977)

[56] T. A. WITTIG: "A Markov Characterization of Generalized Evolution Processes", Preprint

[57] J. ZABCZYK: "Linear Stochastic Systems in Hilbert Space; Structural Properties and Limit Behaviour", Institute of Mathematics, Polish Academy of Sciences, March 1981, Preprint No 236

Paul Krée

MARKOVIANIZATION OF RANDOM VIBRATIONS

1. INTRODUCTION

The infinite dimensional filtering theory is today well developped and important theoretical and practical problems in this aera have been solved : see for example the present volume and [4], [6], [11] .

This lecture is motivated by the constructive study of non linear random vibrations governed by stochastic differential equations (SDE)

$$\dot{\eta}_t = f(\eta_t\ ,\ exc_t) \qquad \text{for}\quad t > 0 \tag{1.1}$$

with a given initial condition for $t = 0$. The driving process (exc_t) called the excitation is usually colored. The uniqueness and the existence for all $t \geqslant 0$ of the solution is assumed : see [10] p.10-14. Currently (exc_t) is a physically realizable and stationary Gaussian process with an irrational spectral density

$$S(\omega) = (2\Pi)^{-1}\,\mathfrak{F}\,C(\omega) = (2\Pi)^{-1}\int_{-\infty}^{+\infty} C(t)\,e^{-it\omega}\,dt \tag{1.2}$$

where C denotes the correlation function In aeronautics for example, exc_t is the vertical component V_z of the turbulent wind V^t and has for spectral density :

$$S_v(\omega) = (a^2 + b^2\ \omega^2)(c^2 + d^2\ \omega^2)^{-11/6} \tag{1.3}$$

with $a, b, c,$ and $d > 0$. In civil engineering the horizontal component V_x of V in the direction 0_x of dominant winds is modelized in any plane $P \perp 0_x$ by a stationary Gaussian process with interspectral measures.

$$S(m,m',\omega) = \omega(1 + k\omega^2)^{-4/3}\ e^{-k'\omega(\alpha\Delta y^2 + \beta\Delta z^2)^{1/2}} \tag{1.4}$$

with $m = (y,z)$ and $m' = (y',z') \in P$, $\Delta y = y-y'$, $\Delta z = z-z'$, k , k' , α , $\beta > 0$.

L. Arnold and P. Kotelenez (eds.), Stochastic Space-Time Models and Limit Theorems, 141–162.

Also for structures submitted to the action of waves, the components of exc_t are built up by integration of some Gaussian random fields on various parts of the structure embedded in the random field : see [13] chapters VI and VIII.

Constructive methods for the study of random vibration use the Markoff property, the Fokker-Planck Equation (PE)..., hence are only valid for Ito's stochastic equations where the driving process is a Gaussian white noise $N_t = \dot{W}_t = dW_t / dt$

Hence these methods are not valid for(1.1). But suppose that a linear and time independent differential filter driven by a white noise

$$\dot{\xi}_t = A\,\xi_t + Q\,\dot{W}_t \tag{1.5}$$

and admitting a stationary solution (ξ_t) can be found such that exc_t is a linear observation of ξ_t where (ξ_t) has no infinite past

$$exc_t = B\,\xi_t \tag{1.6}$$

where B is linear. Then the elimination of exc_t between (I.1) and (I.6) would give an Ito's equation for the augmented process $\zeta_t = (\xi_t , \eta_t)$:

$$\begin{aligned} \dot{\xi}_t &= A\,\xi_t + Q\,\dot{W}_t \\ \dot{\eta}_t &= f(t\,,\,B\,\xi_t) \end{aligned} \quad \Longleftrightarrow \quad \begin{aligned} d\xi &= A\,\xi_t\,dt + Q\,dW \\ d\eta &= f(t\,,\,B\,\xi_t)\,dt \end{aligned} \tag{1.7}$$

Hence the constructive methods can be applied to (I.7). We say that (1.1) has been Markovianized since the augmented process (ζ_t) has the Markoff property. Hence the following filtering problems arise in the study of random vibrations :

(I.8) Problem 1 : For a given stationary and Gaussian physically realizable process exc_t with an irrational spectrum, find a linear differential filter (I.5) admitting a stationary state (ξ_t) such that (1.6) holds for a linear mapping B. What is the class of these linear filters ? This class is well known in the finite dimensional case since then the matrix A is asymptotically stable : $\lambda \in \mathrm{Sp}\,A \Rightarrow \mathrm{Re}\,\lambda < 0$. Note that ξ cannot be a finite dimensional process since then $B\,\xi$ would necessarily have a rational spectral density. Since computers work only with a finite information, the solution of problem 1 is not sufficient and engineers are interested by approximate Markovianization :

(1.9) Problem 2 : Make the Markovianization sufficiently explicit in order to help engineers in the construction of finite dimensional approximations of the infinite dimensional filter (1.5)

(1.10) Problems 3,4 ... Extend the mathematical theories concerning usual Stochastic Differential Equation (SDE) such as P. Malliavin's theory [20] [23] to the infinite dimensional SDE of type (1.7). This needs for example an extension of the usual theory of Sobolev spaces on Gaussian spaces [19] [17] [18] ... to Banach-valued Sobolev spaces with respect to non Gaussian measures.

My contribution to the Bremen workshop was mainly concerned with the solution published in [14] of problems 1 and 2. This solution is presented briefly and in a different way in parts II, III, IV. In particular further examples and references showing the mathematical interest of asymptotically stable linear systems and the practical interest of the approximate Markovianization are given in parts III and IV resp. In order to give more interest to this written version of the lecture, some preliminary results concerning problems 3,4 ... are given in parts V and VI. These new results are obtained using the absolute and the relative differential calculus for measures developed some time ago [17] [18] [19] and applied at that time only to Gaussian measures and to Hilbert-valued functions. Three theorems are proved. The first is of a new kind. The last two theorems reinforce two theorems of P. Malliavin's theory in the following sense : using weaker hypothesis stronger conclusions are proved. Note that Sobolev spaces $W^{p,1}$ of order one with resp to a quasi-invariant and non Gaussian measure M have been studied previously [1] [20] [26]. In the present case M is not necessarely quasi - invariant and all Sobolev spaces $W^{p,k}$ will be defined and used.

2. CYLINDRICAL FORMALISM IN THE LINEAR FILTERING THEORY

The language and the concepts of the elementary linear filtering theory (see for ex [9] chap. V and [13] chap. IV) are weakened in order to obtain a formulation that is also valid in the case of infinite dimensional filters. But there is no miracle : in each given situation the weak results obtained in this way have to be sharpened in a second step in order to eliminate the cylindrical concepts.

This weakening is realized below in the following way :

- a finite dimensional vector space V is viewed as the space $(V')'$ of all linear forms on the dual space V'
- any V-valued random variable ξ is viewed as a cylindrical V-valued random variable i.e. as a linear mapping (linear process)

$$V' \ni u \longrightarrow u \circ \xi = \langle \xi , u \rangle \in L^{o}(\Omega) \tag{2.1}$$

- For any vector space W, a W- valued linear observation $B\xi$ of ξ is identified with the corresponding cylindrical random vector of W i.e. with the product of (2.1) with the transpose B^T of B. Hence :

$$\forall \omega' \in W' \quad \langle B\xi , \omega' \rangle = \langle \xi , B^T \omega' \rangle \tag{2.2}$$

. matrices are identified with the corresponding bilinear forms. For example the familiar definition of the covariance of a V-valued stationary process $q = (q_s)$ on the line

$$C_q(s-t) = E\,[q_s \, q_t^{\,T}] \tag{2.3}$$

with $V \simeq \mathbb{R}^d$ is replaced by the following where $C_q(s-t)$ is now viewed as a bilinear form on $V' \times V'$

$$V \times V' \ni (v,v') \longrightarrow \langle C_q(s-t)v,v' \rangle = E\,[\langle q_s,v' \rangle \langle q_t,v \rangle] \tag{2.4}$$

. Hence for any locally convex Hausdorff space (ℓ.c.H.s) Y a cylindrical random vector of Y' is defined by an element of the space $L(Y,L^2(\Omega))$ of all linear and continuous mappings $Y \to L^2(\Omega)$. Endowing this space with the topology of simple convergence a Y'-valued cylindrical process on the line is defined by a continuous mapping :

$$\mathbb{R} \ni t \longrightarrow \vec{\xi}_t \in L(Y,L^2(\Omega)) \tag{2.5}$$

White noises are defined by generalized cylindrical processes i.e. by linear and continuous mappings

$$\mathcal{D}(\mathbb{R}) = \mathcal{D} \ni \varphi \to \vec{\xi}_\varphi \in L(Y,L^2(\Omega)) \tag{2.6}$$

This can be equivalently defined by linear mappings :

$$\mathcal{D} \otimes Y \ni \varphi \otimes y \longrightarrow \langle \xi_\varphi , y \rangle \in L^2(\Omega) \tag{2.7}$$

separately continuous in theis arguments φ and y

(2.8) <u>Gaussian white noise N with co variance</u>

$$C_N(s-t) = Q\,\delta_o(s-t) \tag{2.9}$$

<u>where $Q \in L_+(Y,Y')$ is given</u>. Let j be the canonical surjection $Y \to Y/\ker Q$. Let j_Q be the product of j with

the canonical injection of Y/Ker Q with its completion for the scalar product defined by Q . Then (Y_Q, j_Q) is called the Hilbert surspace of X defined by Q. Identifying isometrically $X_N = L^2(\mathbb{R}, X_Q)$ with a space of (centered) Gaussian random variables, N is defined as the product of the following two linear mappings

$$\mathcal{D} \otimes Y \xrightarrow{j \otimes j_Q} L^2(\mathbb{R}) \otimes Y_Q \longrightarrow Y_N \qquad (2.10)$$

where j denotes the injection of $\mathcal{D}$ in $L^2(\mathbb{R})$

(2.11) <u>Weak formulation of the elementary linear filtering theory</u>. The familiar definition of the filtering of an $\mathbb{R}^d$-valued white noise N by a causal filter $k \in L^2(\mathbb{R}^+, \mathrm{Mat}(d',d))$:

$$N \longrightarrow (k * N)_t = \int_{-\infty}^{t} k(t-s)\, N_s \, ds \qquad (2.12)$$

has non mathematical meaning ([9] p.284) and [8] gives no meaning to the last integral since $k(t-.) \notin \mathcal{D}$. Following [13], chap.12 a meaning is given to the last integral defining the cylindrical vector associated with the random variable $(k * N)_t$ by

$$(\mathbb{R}^{d'})^* \ni u \to \langle k * N)_t, u\rangle = \int N_s \, k(t-s)^T u \, ds$$

Hence

$$< (k * N)_t \, , u > = < N_. \, , k(t-.)^T u > \qquad (2.13)$$

= random variable associated by N with the element $k(t-.)^T u \in L^2(\mathbb{R}, \mathbb{R}^d)$

The formula (2.13) gives a weak dimension free-formulation of the convolution filtering. In fact let X be a ℓ.c.H.s, a white noise N of type (2.8) and an X-valued causal filter k adapted to N i.e. a linear and continuous mapping

$$X \ni x \longrightarrow k(.)^T x \in L^2(\mathbb{R}^+, Y) \qquad (2.14)$$

Then $k * N$ is a process on the line with values cylindrical vectors of X' , and $k * N$ is defined by (2.13).

Linear filtering is defined in the same way. For example the Wiener-process W of covariance Q min(s,t) on

$\mathbb{R}^+ \times \mathbb{R}^+$ is defined by integration on $\mathbb{R}^+$ of the white noice (2.8). Therefore $\dot{W} = N$ and the right hand side of (2.13) can also be viewed as a Wiener integral [12]

(2.15) Finally the definition of the <u>Reproducing Hilbert space</u> H_Q of a positive quadratic form $\langle Qy, y\rangle$ on some ℓ.c.H.s. Y is recalled : H_Q is the completion of the image of $Q \in L_+(Y,Y')$ for the following scalar product on Im Q :

$$[Qy, Qy'] = \langle Qy, y' \rangle \tag{2.16}$$

(2.17) In general the reproducing space H_Q and the Hilbertian surspace Y_Q of a positive quadratic form $\langle Qy,y\rangle$ on Y are in duality : see [13] § 10.7.

For example if ξ is a centered cylindrical random vector of Y', the covariance of ξ is the following quadratic form

$$\langle Qy, y' \rangle = E[\langle \xi, y\rangle \langle \xi, y'\rangle]$$

and H_Q is called the reproducing Hilbert space of ξ : see [13] § XI.6.

3. STATIONARY GAUSSIAN AND MARKOVIAN CYLINDRICAL PROCESSES

Let X be a ℓ.c.H.s. Let (ξ_t) be a stationnary X'-valued cylindrical Gaussian process with injective covariance C. Below X is Hilbertian since the space X is replaced by its completion for the scalar product $\langle Cx, x'\rangle$. For any s, M_s^- (resp M_s) denotes the closed linear space of $L^2(\Omega)$ generated by $\{\langle \xi_t, x\rangle, t \leqslant s$ and $x \in X\}$ (resp of $\{\langle \xi_s \rangle, x \in X\}$. Denoting Proj_s and Proj_s^- the operators of orthogonal projection in $L^2(\Omega)$ on M_s and M_s^- resp (ξ_t) is called weakly Markovian if

$$x \in X \text{ and } s \leqslant t \Rightarrow \mathrm{Proj}_s^-(\langle \xi_t, x\rangle) = \mathrm{Proj}_s(\langle \xi_t, x\rangle)$$

(3.2) <u>Lemma</u>. For arbitrary $t > 0$ and $x \in X$, let $S_t x$ be the element of X such that for all reals s

(3.3) $\mathrm{Proj}_s(\langle \xi_{s+t}, x\rangle) = \langle \xi_s, S_t x\rangle$

Then S_t is a linear contraction in X

(3.4) $\forall x \in X \quad \| S_t x - x \| \to 0$ if $t \downarrow 0$

and the covariance of ξ is :

$$E\left[<\xi_s, x><\xi_t, y>\right] = \begin{cases} <x, S_{t-s} y> & \text{for } t \geq s \\ <S_{s-t} x, y> & \text{for } s \geq t \end{cases} \qquad (3.5)$$

In fact S_t is contracting for all $t \geq 0$ since

$$\| S_t x \| = \| <\xi_o, S_t x> \| = \| \mathrm{Proj}_o <\xi_t, x> \| \leq \| <\xi_t, x> \| = \| x \|$$

and for $t \downarrow 0$

$$\| S_t x - x \| = \| <\xi_o, S_t x> - <\xi_o, x> \|$$

$$= \| \mathrm{Proj}_o (<\xi_t, x> - <\xi_o, x>) \| \leq \| <\xi_t, x> - <\xi_o, x> \| \to 0$$

(3.6) Proposition. The X'valued stationary cylindrical Gaussian process ξ is weakly Markovian iff $\{S_t, t \geq 0\}$ is a semi-group $\in L(X)$

For the necessity [12] we have denoting $\mathrm{Proj}_{o,s}$ the orthogonal projection on the Gaussian subspace generated by ξ_o and ξ_s in $L^2(\Omega)$

$$<\xi_o, S_{s+t} x> = \mathrm{Proj}_o <\xi_{s+t}, x> = \mathrm{Proj}_o (\mathrm{Proj}_{o,s} <\xi_{s+t}, x>)$$

$$= \mathrm{Proj}_o (\mathrm{Proj}_s <\xi_{s+t}, x>) = \mathrm{Proj}_o <\xi_s, S_t x> = <\xi_o, S_s S_t x>$$

Hence (S_t) is a semi-group. Conversely it (S_t) is a semi-group.

$$\mathrm{Proj}_o \mathrm{Proj}_{o,s} <\xi_{s+t}, x> = \mathrm{Proj}_o (\mathrm{Proj}_s <\xi_{s+t}, x>)$$

Hence $\mathrm{Proj}_{o,s} <\xi_{s+t}, x> = \mathrm{Proj}_s <\xi_{s+t}, x>$

By induction for $s_1 < s_2 < \ldots < s_n < t$

$$\mathrm{Proj}_{s_n} <\xi_t, x> = \mathrm{Proj}_{s_1, s_2, \ldots, s_n} <\xi_t, x>$$

Hence ξ is weakly Markovian and the Proposition is proved

If ξ is weakly Markovian the domain D of the generator A of the semi-group (S_t) is endoved with the graph norm, the canonical injection $i : D \to X$ is continuous. Denoting

A^t the transpose of the linear operator A of X , the transpose of $A \in L(D,X)$ is denoted $A^T_{ex} \in L(X',D')$ and extends A^T

$$D \underset{\overset{i}{\hookrightarrow}}{\xrightarrow{A}} X \simeq X' \underset{\overset{i^T}{\hookrightarrow}}{\xrightarrow{A^T_{ex}}} D' \tag{3.7}$$

(3.8) <u>Proposition</u>. Let ξ be an X'-valued Gaussian and weakly Markovian cylindrical process. Then the D'-valued cylindrical process

$$<W_t, x> = <\xi_t , x> - <\xi_o , x> - \int_o^t <\xi_u , A_x> du \tag{3.9}$$

is a Wiener cylindrical process with covariance $Q \min(s,t)$ on $\mathbb{R}^+$ where

$$Q = i^T \circ A - A^T_{ex} \circ i \tag{3.10}$$

In fact (3.9) defines W as a Gaussian (and centered) cylindrical process. Hence (3.8) is proved simply computing the covariance of W

(3.11) <u>Corollary</u>. The derivation with respect to t of generalized cylindrical processes gives starting from (3.9) :

$$\dot{\xi} = - A^T \xi + \dot{W} \tag{3.12}$$

where the D'-valued cylindrical process $A^T \xi$ is defined by $<A^T \xi , \varphi> = <\xi , A\varphi>$.

Conversely we start with an asymptotically stable linear system i.e. with an Hilbert space X and a continuous semi-group $(\exp tA , t \geqslant 0)$ of contractions in $L(X)$ satisfying the following condition

$$\forall x \in X \quad \| (\exp tA) x \| \to 0 \quad \text{if} \quad t \to \infty \tag{C}$$

Defining Q by (3.10) we are interested by the existence and uniqueness of a D'-valued Gaussian and weakly Markovian cylindrical process ξ solution of (3.12). Introducing also the Hilbert surspace (D_Q, j_Q) of Q we have for arbitrary $x \in D$ [21][14]

$$\int_o^\infty \| j_Q e^{tA} x \|^2 dt = \lim_{T \to \infty} \int_0^T \ldots = \lim (\|x\|^2 - \|e^{tA} x\|^2 = \| x \|^2$$

(3.13) Hence a linear and isometric imbedding

$$X \ni x \xrightarrow{k'} (t \to j_Q \, e_+^{tA} \, x) \in L^2(\mathbb{R}, D_Q)$$

(3.14) Theorem [12][14][21]. Let X be an Hilbert space and let $\{\exp tA, t \geqslant 0\}$ be a continuous semi-group of contractions $\in L(X)$ satisfying (C)
a) Let $N = \dot{W}$ be a Gaussian white noise with covariance $Q\,\delta_o(s-t)$ where Q is given by
(3.10) Then the D'-valued cylindrical process $\xi = k' * N$ is stationary, Gaussian weakly Markovian and satisfies (3.12). Conversely Let ξ be a D'-valued cylindrical stationary Gaussian process ξ without infinite past. i.e. $\cap M_t = \{0\}$ where M_t denotes the closed subspace of $L^2(\Omega)^t$ generated by the variables $\langle \xi_s, x \rangle$, $s \leqslant t$ and $x \in X$. Then $\xi = k' * N$.

(3.15) The two Hilbert spaces defined by a weak asymptotically stable linear filter. Denoting ξ the stationary solution and $C\,\delta_o(s-t)$ the covariance of the corresponding white noise, these two spaces are respectively the reproducing Hilbert space H_ξ of ξ_t and the Hilbert surspace (j_Q, D_Q) of Q.

Since D_Q is the Gaussian space of random variables driving ξ, D_Q is appropriate for the stochastic calculus of variations [22] concerning gentle perturbations of ξ. And since $m = \text{Law}\,\xi_t$ has a good behaviour with resp. to all translations in H_ξ, H_ξ is appropriate for the formulation of the weak differential calculus relatively to m : see part VI.

The particular case where $D_Q = H_\xi$ is the simplest

(3.16) The P. Malliavin's linear filter [22] is defined by an infinite dimensional and separable real Hilbert space X and by the continuous semi-group $\{\text{Id} \exp - t/2, t \geqslant 0\}$ in $L(X)$, where Id denotes the identity mapping of X. In fact, in this case $D = X = D'$ and $Q = \text{Id}$. Hence the two Hilbert spaces H_ξ and D_Q coincide. The solutions of the corresponding stochastic equation

$$\dot{\xi} + \xi/2 = \dot{W} = N \iff d\xi = -\xi dt/2 + dW \tag{3.17}$$

define a Markoff process living in some completion $\tilde{X}$ of X

and the canonical cylindrical measure of X is invariant.

In order to build up a process of the same nature but living in X , the drift term $-\xi\, dt$ in (3.17) can be replaced by a stronger one :

(3.18) The B. Gaveau's linear filter. Let u be a symmetric injective Hilbert Schmidt operator of a separable Hilbert space X . Let (e_n) be an orthonormal basis of X .

$$u(e_n) = \lambda_n^{-1/2} e_n \; ; \; \Sigma \lambda_n^{-1} < \infty$$

Hence $X \sim \ell_2 = \{x = \Sigma x'_n e_n \; ; \; \Sigma x'^2_n < \infty\}$

In [7] the following X-valued stochastic equation is defined

$$d\, \xi_n = - \lambda_n \xi_n/2 + d\, W_n \qquad n = 1,2 \tag{3.19}$$

where the Wiener processes W_n are independent

The process $\xi = \xi_n)$ admits the invariant measure

$$\gamma = \prod_{n=1}^{\infty} \gamma_n \quad \text{with} \quad \gamma_n = \left(\frac{\lambda_n}{2\Pi}\right)^{1/2} \left(\exp - \frac{\lambda_n}{2} x'^2_n\right) dx'_n$$

Let H_R be the Hilbert space admitting the $\varepsilon_n = \lambda_n^{-1/2} e_n$ as an orthonormal basis :

$$H_R = \{ \Sigma_1^{\infty} x_n \varepsilon_n \; ; \quad \Sigma x_n^2 < \infty\}$$

Putting

$$x'_n = \lambda_n^{-1/2} x_n \tag{3.20}$$

the canonical injection of H_R in X :

$$H_R \ni (x_n) \xrightarrow{\;i\;} x' = (x'_n) \in X$$

is Hilbert-Schmidt and maps the canonical cylindrical measure of H_R on the Radon measure γ . This means that H_R is the reproducing Hilbert space of γ . The Gaussian Hilbert space defined by the driving process (W_n) is $X \neq H_R$. Therefore the coordinate transformation (3.20) permits the conversion of the usual formulation (with x_n coordinates) of results concerning the differential calculus connected with γ to a formulation compatible with the stochastic calculus of variation (using x'_n coordinates).

4. APPLICATION TO MARKOVIANIZATION [14]

The method is explained below. Let us consider a random vibration driven by a scalar physically realizable process

$$exc_t = k * N \qquad (4.1)$$

with $k \in X = L^2(\mathbb{R}^+)$ and where N denotes a stationary scalar Gaussian white noise with covariance $\delta_o(s-t)$. For example $exc_t = V$ with spectral measure (1.3), the Laplace transform of k is

$$p \to \int_0^\infty e^{-pt}\, k(t)\, dt = \frac{a + bp}{(c + dp)^{11/6}}$$

The following semi-group in $L(X)$

$$\exp tA : f \to \tau_t f = f(t + .) \ ; \quad t > 0 \qquad (4.2)$$

is a continuous semi-group of contractions in $L(X)$ satisfying the condition (C) .

(4.3) <u>The corresponding weak asymptotically stable linear filter</u> is constructed using the procedure (3.7). Hence the domain D of $A = d/dt$ in the Sobolev space $H^1(\mathbb{R})$. The dual D' is the subspace $\overset{o}{H}{}^{-1}(\mathbb{R}^+)$ of elements $u \in H^{-1}(\mathbb{R})$ supported by $[0 , + \infty[$. Hence Q is $\varphi \to \varphi(0)\, \delta_o$. Therefore D_Q is $\mathbb{R}$ and j_Q is

$$D_Q \ni \varphi \xrightarrow{j_Q} \varphi(0) \in \mathbb{R} \qquad (4.4)$$

The dynamic equation of this filter is (1.5) where $A^T = - d/dt$ and where N has covariance $Q\, \delta_o(s-t)$

(4.5) <u>Proposition</u> [14]. For any causal filter $k \in X$, the physically realizable process $(k * N)_t$ coincides with the linear observation $B\,\xi_t$ of the stationary state ξ of (I.5) where B denotes the linear form on X defined by k :

$$B\,\xi_t = \langle \xi_t , k \rangle \qquad (4.6)$$

The brackets have a meaning since k belongs to the Reproducing Hilbert space of ξ_t . Note that the complete definition of the Markovianization $\zeta = (\xi,\eta)$ of (1.1) needs the defintion of all probalities of transition, hence the introduction of some completion $\hat{X}$ of X where the random vectors ξ_t take their values.

Proof of (4.5). According to the definition (3.12) of ξ by filtering N :

$$B\,\xi_t = <\xi_t\ ,\ k> = <N_u\ ,\ j_Q\ e^{(t-u)A}\ k>$$

Hence using (4.2)

$$= <N_u\ ,\ j_Q(k(t-u+.))> = <N_u\ ,\ k(t-u)> = (k * N)_t$$

Note that the driving white noise N is identical with the white noise expressing exc_t in (4.1). Hence the Markovianization does not need an enlarging of the filtration

(4.7) Reduction of the Markovianization :
In the previous construction , $X = L^2(\mathbb{R}^+)$ can be replaced by the closed subspace X_h of X generated by all right translated functions $h(t+.)$ of h . For example, if $N = \dim X_h < \infty$, the vibration driven by h can be Markovianized by simply adding N components to ξ .

The Markovianization of vibrations driven by a vectorial process or by a physically realizable random field uses the same method : see [14].

(4.8) Approximate Markovianization :
Denoting L_ℓ the ℓth Laguerre polynomial and putting

$$e_\ell = (\ell!)^{-1}\ L_\ell(x)\ \exp(-x/2)$$

then $\{e_\ell\ ,\ \ell = 0,1\ \ldots\}$ is an orthonormal basis of X such that for any M , the subspace X_M of X generated by $e_o, e_1, \ldots e_M$ is invariant by all left translation. Hence in view of (4.7) an approximate Markovianization of (1.1) driven by $exc_t = k * N$ is constructed proceeding as follows :

. The element $k \in X$ is expanded in term's of the e_ℓ :

$$k \sim \Sigma_o^M\ k_\ell\ e_\ell \qquad\qquad k_\ell = <k\ ,\ e_\ell>$$

. The auxiliary dynamic system is defined by the restriction of $\{\exp At\}$ to the subspace X_M of X

(4.9) Some recent progress concerning approximate Markovianization.
References [2] [3] give applications for simulation procedu-

res. Reference [27] refers to numerical algorithms concerning fatigue loads of off shore structures.

5. <u>THE WEAK ABSOLUTE DIFFERENTIAL CALCULUS FOR MEASURES.</u>
Let X be a ℓ.c.H.s. The family $\{X'_\alpha, \alpha \in A\}$ of all finite dimensional subspaces of the dual space X' is directed for the ordering defined by the inclusion relation. The notation $X'_\alpha \Subset X'$ means that X'_α is a finite dimensional subspace of X'. Hence an inductive system $\{X'_\alpha, i_{\alpha\beta}\}$ of finite dimensional vector spaces. Hence transposing and putting $X_\alpha = (X'_\alpha)' \simeq X/(X'^{\perp}_\alpha)$ an inductive system $\{X_\alpha, s_{\alpha\beta}\}$ of vector spaces and a lifting (X, s_α) of this system are constructed Hence a diagram

$$\begin{array}{ccccc} & & & & V \\ & \overset{f}{\nearrow} & & & \uparrow f_\beta \\ X & \xrightarrow{s_\alpha} & X_\alpha & \xrightarrow{s_{\alpha\beta}} & X_\beta \\ \vdots & & \vdots & & \vdots \\ X' & \xleftarrow{i_\alpha} & X'_\alpha & \xleftarrow{i_{\alpha\beta}} & X'_\beta \end{array} \tag{5.1}$$

where doted lines represent dualities. For any ℓ.c.H.s. V a V-valued function f on X is called cylindrical if $f = s_\beta \circ f_\beta$ for some $\beta \in A$ and some $f_\beta : X_\beta \to V$ and X_β is called a basis of f. The space $C^\infty_o(X_\alpha)$ of C^α functions φ on X_α such that for any fixed j, $\nabla^j \varphi(x) \to 0$ if $|x| \to \infty$ has a natural Frechet topology. The Hahn Banach theorem shows that the dual of $\mathcal{T}(X_\alpha) = O_c(X_\alpha)$ $= \lim\limits_{\to k} (1 + |x|^2)^k C^\infty_o(X_\alpha)$ is the L. Schwartz $0'_c(X_\alpha)$ of all convolution operators in $\mathcal{S}(X_\alpha)$. For $\beta \leqslant \alpha$ the mapping

$$\mathcal{T}(X_\beta) \ni F \longrightarrow F \circ s_{\alpha\beta} \in \mathcal{T}(X_\alpha) \tag{5.2}$$

identifies $\mathcal{T}(X_\beta)$ to a topological subspace of $\mathcal{T}(X_\alpha)$.
A ℓ.c.H.s. of cylindrical test functions on X is defined putting

$$\mathcal{T}_{cyl}(X) = \lim_{\to} \mathcal{T}(X_\alpha) \tag{5.3}$$

Following [17] [19], an element T of the space $L(\mathcal{T}_{cyl}(X), V)$ is called a V-valued <u>cylindrical distribution</u> on X. Hence T is characterized by a coherent family of V-valued distributions T_α on the spaces X_α.

(5.4) Let M denote an interiorely regular probability measure on the Baire σ-field of X s.t. $\mathcal{C}_{cyl}(X) \subset \cap L^p_M(X)$. Hence M is characterized by the cylindrical distribution (M_α) on X with $M_\alpha = (s_\alpha)_*(M)$. We also assume that the mappings s_α $(\alpha \in A)^\alpha$ generate the completed σ-field of X.

(5.5) Martingale representation of any $F \in L^p_M(X,V)$ for $1 < p < \infty$ and for a Banach space V with the Radon - Nikodym property. In fact, F is characterized by the cylindrical distribution $(F_\alpha M_\alpha)$ where the L^p-norms of the F_α $= E[F \| s_\alpha]$ are uniformly bounded [17] [19].

(5.6) Weak derivative $\nabla^k M$ for any given $M \in L(\mathcal{C}_{cyl}(X),V)$. For any integer $k > 0$ a multilinear and symmetric mapping :

$$X^k \ni (x_1,\ldots,x_n) \to \partial_{x_1 x_2 \ldots x_k} M \in L(\mathcal{C}_{cyl}(X),V) \tag{5.7}$$

is defined recursively putting

$$< \partial_x M \, , \, F > = - < M \, , \, \partial_x F > \tag{5.8}$$

As in [17] [19] the global weak derivative $\nabla^k M$ of order k is defined by the linear mapping

$$\odot_k X \xrightarrow{\nabla^k M} L(\mathcal{C}_{cyl}(X) \, , \, V) \tag{5.9}$$

defined by (5.7) or equivalently by the corresponding linear mapping

$$\mathcal{C}_{cyl}(X) \xrightarrow{\nabla^k M} \mathrm{Mul}_{sym}(X^k \, , \, V) \tag{5.10}$$

Hence $\nabla^k M$ is characterized by the coherent family of derivatives $\nabla^k M_\alpha : \mathcal{C}(X_\alpha) \to \mathrm{Mult}_{sym}(X_k,V)$ of the measures M_α

(5.11) H-smoothness of probability measures on X

Let H be a ℓ.c.H.s. continuously and linearly injected in X. A probability M on X is called k-times H-derivable if M satisfies (5.4) and if for arbitrary $j \leqslant k$ and $1 < p \leqslant \infty$, the linear mapping $\mathcal{C}_{cyl}(X) \to \mathrm{Mul}_{sym}(H^j,\mathbb{R})$ admits a continuous extension :

$$\nabla^j M \in L(L^p_M(X) \, , \, \mathrm{Mul}_{sym}(H^j,\mathbb{R})) \tag{5.12}$$

Usually the vectorial measure $\nabla^j M$ is absolutely continuous with respect to M i.e.

$$\nabla^j M = C_j M \tag{5.13}$$

where the function C_j takes values in some completion of $Mul_{sym}(H^j, \mathbb{R})$. In the same way :

$$\nabla^j M_\alpha = C_{j,\alpha} M_\alpha \quad \text{with} \quad M_\alpha \in L^{p'}_{M(\alpha)}(X, \odot_j X_\alpha) \tag{5.14}$$

Hence for arbitrary $h \in H$

$$\nabla^j M \cdot h^j = (\partial_h)^j M = C_j(x).h^j M(dx) \tag{5.15}$$

(5.16) Remarks. a) If M is k-times H-derivable for $k = 1,2 \ldots$ then M is not necessarely H-quasi-invariant : see for example on $X = H = \mathbb{R}$ the C^∞ measure $M(x)$ vanishing for $x < 0$ and $= \exp -x^{-1}$ for $0 < x < 1$
b) But if M is H-quasi-invariant and H-C1 , our definition of ∇M is coherent with the usual one using translations. In fact for arbitrary $h \in H$ and $\lambda \to 0$

$$\partial_h M = \lim \lambda^{-1}(\tau_{\lambda h} M - M) \quad \text{in } \mathcal{C}_{cyl}(X)' \text{ weak}$$

(5.17) Theorem 1. With the previous notations let M be an H-C^∞ probability measure one X such that for arbitrary $p \in]1, +\infty[$

$$\sum_{\ell=0}^{\infty} \ell!^{-1} \| C_\ell(x) \cdot h^\ell \|_p < \infty \tag{5.18}$$

and such that the cylindrical polynomials are dense in L^p. Then M is quasi-invariant with respect to all translations in H and

$$\frac{d(\tau_h M)}{dM} = \sum_{\ell=0}^{\infty} \frac{C_\ell(x)\, h^\ell}{\ell!} \in \cap L^P_M(X) \tag{5.19}$$

Proof. This sum converges in L^p by (5.18). Moreover for the arbitrary cylindriacl polynomial function F

$$< M , \tau_h F > = \int M(dx)\, F(x-h)$$
$$= \sum_{\ell=0}^{\infty}(-1)^\ell < \partial_h^\ell F , M > \ell!^{-1} = \sum_{\ell=0}^{\infty} < F , \partial_h^\ell M >$$
$$= \int F(x) \left[\sum_{\ell=0}^{\infty} \ell!^{-1} C_\ell(x) \cdot h^\ell\right] M(dx) = < \tau_{-h} M , F >$$

(5.20) Weak divergence of cylindrical distribution. Putting $\mathcal{C}(X_\alpha, k) = \mathcal{C}(X_\alpha, \odot_k X'_\alpha)$ we have for

$X'_\beta \Subset X'_\alpha \Subset X'$ a canonical injection

$\mathcal{C}(X_\beta , k) \ni F \to (\odot_k \ i_{\alpha\beta}) \ F \ (s_{\alpha\beta}) \in \mathcal{C}(X_\alpha ,k)$

Hence a space of vectorial and cylindrical test functions is defined on X putting

$$\mathcal{C}_{cyl}(X , k) = \lim_{\rightarrow} \mathcal{C}(X_\alpha , k) \tag{5.21}$$

The weak divergence

$$L(\mathcal{C}_{cyl}(X , k+1) , V) \ni \to \delta \ T \in L \ (\mathcal{C}_{cyl}(X , k) , V) \tag{5.22}$$

has been defined in [17-19] by

$$< \delta \ T , F > = < T , \nabla F > \tag{5.23}$$

(5.24) Proposition. For any $G \in \mathcal{C}_{cyl}(X)$ and for any cylindricial measure $M = (M_\alpha)$ on X satisfying (5.4) and cylindrically smooth

$$\nabla^k (G \ M) \ . \ y^k = (\ B_k(G) \ . \ y^k) \ M \tag{5.25}$$

with $B_k(G) \ . \ y^k = \sum_{\ell=0}^k \binom{k}{\ell} (\nabla^{k-\ell} G.y^{k-\ell})(C_\ell(x).y^\ell)$

Hence putting $\partial^k_y = (\partial_y)^k$ the following duality formula with $F \in \mathcal{C}_{cyl}(X) , y \in X$

$$\int G(\partial^k_y F) \ M = < F , (B_k(G) \ . \ y^k)M > \tag{5.26}$$

In the same way for arbitrary $Q \in \mathcal{C}_{cyl}(X,k')$ and $G \in \mathcal{C}_{cyl}(X)$

$$\delta^k (Q \ G) = \sum_{\ell=0}^k \binom{k}{\ell} (\delta^\ell \ Q) \ . \ (\nabla^{k-\ell} \ G) \tag{5.27}$$

The proof uses the following formula valid for arbitrary F and $G \in \mathcal{C}_{cyl}(X)$

$$G(\partial^k_y \ F) = \sum_{\ell=0}^k (-1)^{k-\ell} \binom{k}{\ell} \partial^\ell_y \ (F \ \partial^{k-\ell}_y \ G) \tag{5.28}$$

This formula is proved by induction on k.

Proof of (5.25) and (5.26) :

$$((-1)^k < F , \partial_y^k(G\, M) > = \int G(\partial_y^k F)\, M$$

$$= \sum_{\ell=0}^k (-1)^{k-\ell} \binom{k}{\ell} \int \partial_y^\ell (F(\partial_y^{k-\ell} G))\, M$$

$$= (-1)^k \sum_{\ell=0}^k \binom{k}{\ell} F(\partial_y^{k-\ell} G)(\partial_t^\ell M) = (-1)^k < F , (B_k(G) . y^k)M >$$

and (5.27) is proved in the same way.

4. THE WEAK RELATIVE DIFFERENTIAL CALCULUS FOR MEASURES. Application to the stochastic calculus :

Let H be a Hilbert subspace of a ℓ.c.H.s. X
The weak differential calculus relative to a given $H-C^\infty$ measure M on X is briefly presented below.

The image of the injective mapping

$$L_M^p(X) \ni F \longrightarrow (G \to \int F\, G\, M) \in \mathcal{C}_{cyl}(X)' \qquad (6.1)$$

is denoted $L_M^P(X)\, M$. In the same way $\mathcal{C}_{cyl}(X)\, M$ can be defined; hence a triplet

$$\mathcal{C}_{cyl}(X)\, M \xrightarrow{i} L^2(X)\, M \simeq (L^2(X)\, M)' \xrightarrow{i^T} \mathcal{C}_{cyl}(X)' \qquad (6.2)$$

(6.3) Definition of the relative weak derivative $(\nabla^k F)\, M$ for any $F \in \cup L_M^P(X)$.

Since M is $H-C^\infty$, the following mapping is continuous for any fixed h .

$$\mathcal{C}_{cyl}(X) \ni G \xrightarrow{B_k(.).h^k} B_k(G).h^k \in \cap L^P(X)\, M \qquad (6.4)$$

Hence the duality formula (5.25) shows that the relative derivation

$$\mathcal{C}_{cyl}(X)\, M \ni F \longrightarrow (\partial_h^k F)\, M \in \mathcal{C}_{cyl}(X)\, M \qquad (6.5)$$

is extended by the transpose of (6.4) :

$$\bigcup_{p>1} L_M^P(X)\, M \xrightarrow{\partial_y^k} \mathcal{C}_{cyl}(X)' \qquad (6.6)$$

More generally, for any $T \in L(L_M^P(X)\, M , V)$,

the relative derivative $\partial_h^k T \in L(\mathcal{C}_{cyl}(X), V)$,

can be defined by

$$(\partial_h^k T)(F) = T(B_k(F).h^k) \tag{6.7}$$

The polynomial function $h \to \partial_h^k T$ is the global relative weak derivative of the cylindrical distribution T

(6.8) Using these preliminaries, Banach-valued Sobolev spaces with resp to the H-smoth measure M are defined as Hilbert valued Sobolev spaces have been defined in [17] [19] for the Gaussian case. As in [17] the study of the density of the subspace $\mathcal{C}_{cyl}$ (: $W^{p.k}$ is reduced to the finite dimensional case using (5.5), and many different kinds of Sobolev spaces can be defined.

(6.9) Note that the Gaussian case is <u>very particular</u> since for any $X'_\alpha \subset X$, M is the product of two measures. Therefore $B_k(.)y^k$ defines a linear and continuous mapping in $\mathcal{C}_{cyl}(X)$ <u>for arbitrary $y \in X$</u>. Hence for any cylindrical distribution $T \in L(\mathcal{C}_{cyl}(X), V)$, the relative weak derivative $\partial_y^k T$ can be defined in the Gaussian case.

(6.10) <u>Theorem 2</u>. (see [23] if g and g' are Gaussian) Let X and Y be two ℓ.c.H.s. Let H(resp K) be a dense Hilbert subspace of X(resp Y) Let g and g' be a H-smooth probability on X and a K-smooth probability on Y s.t. cylindrical test functions are dense in the Sobolev spaces with respect to g and g' resp. a) Then $gg' = g \otimes g'$ is $H \oplus K$-smooth. b) Moreover, for arbitrary $p>2$, k and $\ell>0$ and for any Hilbert space G the following mapping is continuous and has norm $\leqslant (k+1)^{1/p}$

$$W^{p,k+\ell}(X \times Y, G) \ni F \longrightarrow \vec{F} = F(x,.) \in W^{p,k}(X, W^{2,\ell}(Y,G)) \tag{6.11}$$

The proof of a) uses an argument of topological tensor product. The proof of b) uses simply the Jensen inequality since putting $H'^i = \hat{\odot}_i H'$ for $i \leqslant k$:

$$(\| \nabla^i \vec{F} \|_p)^p = \int_X g(dx) \Big(\sum_{j=0}^{\ell} \int_Y \left| \nabla_x^i \nabla_y^j F(x,y) \right|^2_{H'^i \otimes K'^j \otimes G} g'(dy) \Big)^{p/2}$$

$$\leqslant \sum_{j=0}^{\ell} \int g(dx) \int \left| \nabla_x^i \nabla_y^j F(x,y) \right|^p_{H'^i \otimes K'^j \otimes G} g'(dy)$$

$$\leqslant \sum_{j=0}^{\ell} \int g(dx) \int \left|\nabla^{i+j} F(x,y)\right|^{p}_{(H+K)',(i+j)\otimes G} g((dy)$$

$$\leqslant (\| F \|_{p,k+\ell})^{p}$$

(6.12) Now a $H-C^{\infty}$ probability measure P on X is considered s.t.

a) $\mathcal{C}_{cyl}(X)$ is dense in $W^{p,k}(X)$ for arbitrary $k > 0$ and $1 < p < \infty$

b) The divergence δ is continuous in the Hilbertian Sobolev spaces relative to P (see [15] in the Gaussian case; this is also a corollary of [24]

(6.13) Therefore a) by the Hölder inequality the Frechet space

$$W = \cap_{p,k} W^{p,k} (X)$$

is an algebra, b) in view of (6.12), (5.27) can be extended by continuity for $G \in W$ and $Q = F P$ with F in some vectorial Sobolev space : see [5] in the Gaussian case.

(6.14) <u>Theorem 3</u>. An Euclidean space Y is identified with the dual space and let ρ be the Riesz isometry of H' on the dual space $H'' = H$. Let $F \in W \otimes Y$ be such that ∇F is a.s. surjective $H \to Y$ and such that :

$$\gamma = \sigma^{-1} \in \cap_p L^p(X, \text{End } Y) \quad \text{with} \quad \sigma F = \nabla F \circ \rho \circ (\nabla F)^T$$

Then denoting m the law of F and putting $G^F = E[G \| F]$ the following linear mapping is continuous

$$W \ni G \longrightarrow G^F m \in \mathcal{S}(Y)dy \tag{6.15}$$

(6.16) <u>Corollaries</u> : a) This gives in particular $m \in \mathcal{S}(Y)dy$ for $G = 1$: compare with [22] and see [16] in the Gaussian case. b) The transpose of (6.15) is the continuous mapping

$$S'(Y) \ni T \longrightarrow T \circ F \in W' \tag{6.16}$$

(see [28] in the Gaussian case)

Proof. Let $(\varepsilon_i) = (\partial_i) = (\partial/\partial y_i)$ be an orthonormal basis of Y. Let $\mathrm{Diff}_p(Y)$ be the algebra of linear differential operators with polynomial coefficients on Y. A lifting of $d \in \mathrm{Diff}_p(Y)$ by F is defined as any $D \in \mathrm{End}\, W$ such that :

$$E[(d\varphi) \circ F \,.\, G] = E[(\varphi \circ F)\, D\, G]$$

for arbitrary $\varphi \in \mathcal{D} = \mathcal{D}(Y)$ and $G \in W$.

Denoting q the weight $1 + |y|^2$ on Y the following fact is well known for any probability measure m' on Y

$$\forall\, \ell \quad \forall\, \alpha = (\alpha_1, \ldots \alpha_n) \quad q^\ell\, \partial^\alpha\, m' \text{ is a bd measure} \Rightarrow m' \in \mathcal{S}\, dy$$

This can be applied to $m' = G^F\, m$. Since the conditionning induces a contraction in all L^p, the problem is to lift on X the differential operators : $\varphi \to d\varphi = \partial^\alpha (q^\ell \varphi)$ on Y. But the operator of product by q^ℓ on Y is lifted in the operator of product by $q^\ell(F) = q^\ell \circ F$ on X. And ∂_i is lifted in

$$G \longrightarrow D_i\, G = \delta((\gamma\, \varepsilon_i \circ \nabla F)\, G)$$

since [22]

$$E[(\partial_i \varphi) \circ F \,.\, G] = E[\langle \nabla(\varphi \circ F), ((\gamma\, \varepsilon_i) \circ \nabla F)\, G]$$

Therefore $d = (\partial_1)^{\alpha_1} (\partial_n)^{\alpha_n} (q^\ell .)$ is lifted in

$$D = (q^\ell(F).) \,.\, (D_n^{\alpha_n}) \ldots (D_1^{\alpha_1})$$

and the theorem 3 is proved.

Paul Krée Mathématiques

et LA n° 213 - Université Paris VI.

Place Jussieu - 75005 PARIS

REFERENCES.

[1] S. Albeverio and R. Høegh-Krohn, Z. Wahrsch-Verw. Gebiete 40 (1977) pp.1-57.

[2] P. Bernard, M. Fogli et C. Wagner. Actes des journées (Juin 1984) de mécanique aléatoire appliquées à la construction. Edited by AFREM and the Lab. Cent. des Ponts et Ch. (Paris) pp.157-165.

[3] P. Bernard, P. Fogli, M. Bressolette, P. Lemaire To appear in Journal de Méc. Théor. et Appl.

[4] R.F. Curtain and A.J. Pritchard Lect. Notes in Control and Inf. Science vol.8 (1978) Springer Verlag.

[5] J. Diebolt. C.R. Acad. Sc. Paris t.296 (Juin 1983) Serie I. PP.837-840.

[6] W.H. Fleming and L.G. Gorostiza Editors : Lecture Notes in Control and Information Sciences n°42. Springer Verlag 1982.

[7] B. Gaveau. GR. Acad. Sc. Paris t.293 (Novembre 1981) Serie I. pp.469-472.

[8] I.M. Gelfand and N.Y. Vilenkin. Les distributions, tome 4. Dunod Paris (1967).

[9] J.L. Guilkman et A. Skorokhod. Introduction à la théorie des processus aléatoires. Ed. MIR Moscou. 1977-1980.

[10] R.Z. Hasminskii. Stochastic stability of differential equations. Sijthoff and Noordhoff (1980).

[11] M. Hazewinkel and J.C. Willems. Editors : Stochastic Systems ... Reidel (1981).

[12] K. Ito in Stochastic Analysis. North Holland 1984, pp.197-224.

[13] P. Krée and C. Soize. Mécanique aléatoire. Dunod Paris (1983) English translation in course by D. Reidel.

[14] P. Krée. Journ. of Math. Phys. 24(11) Novembre 1983. pp.2573-2580.

[15] M. and P. Krée. C.R. Acad. Sc. Paris t296 (Juin 1983) Serie I. pp.833-836.

[16] P. Krée. C.R. Acad. Sc. Paris. t.296 (Janvier 1983) Serie I. pp.223-225.

[17] P. Krée in Sem. P. Lelong I Lecture Notes in Mathematics (Springer Verlag) n°410 (1973). and II Lecture Notes in Mathematics n°474 (1974) pp.2.47

[18] P. Krée Journ. of Funct. Anal. Vol.31 n°2 (1979) pp.150-186

[19] P. Krée.Séminaire sur les ed.p en dim. infinie 1974-1975. Edited by Secrétariat math. of the H. Poincaré Institute (Paris)

[20] S. Kusuoka. J. Fac. Sci. Univ. Tokyo 29 (1982) pp.79-85.

[21] J.T. Lewis and L.C. Thomas. Wahrsch. Verw. Geb. 30, pp.45-55 (1974).

[22] P . Malliavin in Proc. Int. Symp. on SDE (1976) Kyoto ed. bt K. Ito Konokoniya, Tokyo (1978)

[23] P. Malliavin in Stochastic Analysis ed. by K. Ito North Holland (1984) pp.369-386.

[24] P.A. Meyer. Manuscript received by P. Malliavin in may 1983 and published in Sem. of Probability XVIII Lecture Notes n° 1059 pp. 179-193. Springer Verlag (1984).

[25] P.A. Meyer. The "Malliavin Calculus" and some pedagogy. to appear in Sem of Probability (Lect. Notes).

[26] P. Paclet. Expose 5 in Sém. P. Krée. 1977-1978. Edited by Secrétariat Math. of the H. Poincaré Institute (Paris).

[27] C. Soize. Actes des journées (Juin 1984, Paris) de mécanique aléatoire appliquée à la Construction. Edited by AFREM and the Lab. Cent. des Ponts et Ch. (Paris) pp.24-31.

[28] S. Watanabe. In Lect. Notes in Control and Inf. Science n°49 (1983) p.284-290, Springer Verlag. Berlin

A. S. USTUNEL

STOCHASTIC ANALYSIS ON NUCLEAR SPACES AND ITS APPLICATIONS

Foreword

In this paper we review the recent developments in the theory of stochastic processes with values in the nuclear spaces. The choice of the subjects follows our personal research and all the results are announced without any proof. We have avoided to give the detailed applications and tried to give an overview of the subject in such a way that the reader can use these results as soon as he becomes familiar with the concepts illustrated here.

Having explained why one needs a stochastic calculus in the frame of the distributions, we have given the basic definitions and some important identification results in the first section. The second section is devoted to the construction of the stochastic calculus and some of its applications. In the third section we study in more detail the trajectories of the weakly regular processes and announce some hypoellipticity results for the stochastic partial differential operators operating on the distributions-valued semimartingales. The weak convergence of the processes and the characterization of the additive processes are also given in this section.

As a conclusion all the results confirm the following (probably hasty) conjecture: any proposition which is true in finite dimension and announcable on a nuclear space is true also there.

L. Arnold and P. Kotelenez (eds.), Stochastic Space-Time Models and Limit Theorems, 163–177.

Motivations

Let W be an $\mathbb{R}^d$-valued standard Wiener process, $T \in \mathcal{D}'(\mathbb{R}^d)$. Define the function f on $\mathbb{R}^d$ by

$$f(\lambda) = \langle T, \varphi(\cdot + \lambda)\rangle, \quad \varphi \in \mathcal{D}(\mathbb{R}^d).$$

Since f is a C^∞-function, Itô's formula gives

$$f(W_t) = T(\varphi) + \int_0^t \nabla f(W_s).dW_s + \frac{1}{2}\int_0^t \Delta f(W_s)\,ds$$

$$= T(\varphi) + \int_0^t \langle T, \partial_i \varphi(\cdot + W_s)\rangle\, dW_s^i + \frac{1}{2}\int_0^t \langle T, \Delta\varphi(\cdot + W_s)\rangle\, ds$$

Let us denote by $\tilde{X}(\varphi)_t$ the process $f(W_t)$. Then $\tilde{X}$ defines a linear mapping from $\mathcal{D}(\mathbb{R}^d)$ into the space of the semimartingales (with respect to some filtration) such that

$$d\tilde{X}(\varphi)_t = 1/2\; \tilde{X}(\Delta\varphi)_t\, dt + \tilde{X}(\partial_i\varphi)_t\, dW_t^i$$
$$\tilde{X}(\varphi)_0 = T(\varphi)$$

for any $\varphi \in \mathcal{D}(\mathbb{R}^d)$. A moment's reflexion suggests that there exists some $X : \mathbb{R}_+ \times \Omega \mapsto \mathcal{D}'(\mathbb{R}^d)$ such that X has continuous trajectories and $\langle X_\cdot, \varphi\rangle = \tilde{X}(\varphi)_\cdot$ for any $\varphi \in \mathcal{D}(\mathbb{R}^d)$. Hence X is a "solution" of

(1) "$dX_t = 1/2\; \Delta X_t\, dt - \partial_i X_t\, dW_t^i$"
$X_0 = T$

where the the integrals are to be defined. In fact what we have done can be explained easily using a physical language: we have two reference frames R1 and R2 such that R2 is moving with respect to R1 following the Brownian path $W_\cdot(\omega)$:

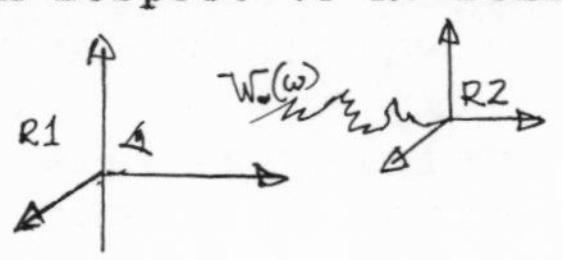

and $T \in \mathcal{D}_2'(\mathbb{R}^d)$ is observed as $X_t(\varphi)$ by an observer in R1 at time t. Since $x \mapsto x+W_t(\omega)$ is reversible, the equation (1) should have a "unique solution" and to prove this, it is sufficient to show that

$$" X_t * \delta_{-W_t} = T "$$

where $\delta_{\cdot}$ is the Dirac measure whose support is "." and "$*$" denotes the convolution. Of course, in order that this discussion to be of some interest we should define first what we understand by the words "solution", "unique", etc. In fact, all we did above can be rigourously explained using the weak formalism except the calculation of the convolution: for this, it is obvious that we need a kind of integration by parts formula. This will be done by defining a class of processes with values in $\mathcal{D}'(\mathbb{R}^d)$ with which it is possible to construct a stochastic calculus. For the sake of generality, instead of working on $\mathcal{D}$ and $\mathcal{D}'$, we will work on the more general spaces, called nuclear spaces.

I. Notations and preliminaries

Φ denotes a complete, nuclear space whose continuous dual Φ' is also nuclear under the strong topology $\beta(\Phi',\Phi)$, denoted by Φ'_β. If U is an absolutely convex neighborhood (of zero) in Φ, we denote by $\Phi(U)$ the completion of $\Phi/p_U^{-1}(0)$ with respect to the norm p_U, where p_U denotes the gauge function of U; k(U) denotes the canonical mapping from Φ into $\Phi(U)$. If $V \subset U$ is another such neighborhood we define $k(U,V)\colon \Phi(V) \longrightarrow \Phi(U)$ as $k(U) = k(U,V) \circ k(V)$.
Let us recall that in each nuclear space Φ, there exists a neighborhood base $\mathcal{U}_h(\Phi)$ such that (cf. [10]) for any $U \in \mathcal{U}_h(\Phi)$, $\Phi(U)$ is a separable Hilbert space and there exists $V \in \mathcal{U}_h(\Phi)$, $V \subset U$ such that $k(U,V)$ is a Hilbert-Schmidt (or 2-nuclear) mapping. If B is an absolutely convex, bounded, closed set, $\Phi[B]$ denotes the subspace spanned by B equipped with the norm p_B (i.e., the gauge function of B).
In this way, the continuous dual of $\Phi(U)$ can be identified as $\Phi'[U^\circ]$ where U° is the polar of U (cf. [10]).

In the sequel we denote by $(\Omega, \mathcal{F}, \mathcal{F}_{t\geq 0}, P)$ a general probability space where $(\mathcal{F}_t, t\geq 0)$ is a right continuous, increasing filtration and $\mathcal{F}_0$ contains all the P-null sets.

Definition 1

i) Let X be the set $\{X^U;\ U\in \mathcal{U}_h(\bar{\Phi}'_\beta)\}$ where, for any U, X^U is a stochastic process with values in $\Phi'(U)$. X is called a projective system (of stochastic processes) if, for any $V\in \mathcal{U}_h(\bar{\Phi}'_\beta)$, $V\subset U$, $k(U,V)\circ X^V$ and X^U are undistinguishable.

ii) X is called a (projective or generalized or g-) semi-martingale, martingale, etc., if for any $U\in \mathcal{U}_h(\bar{\Phi}'_\beta)$, X^U is respectively a semimartingale, martingale, etc., with values in $\bar{\Phi}'(U)$. More generally we say that X possesses the property π if X^U possesses the property π for any U in $\mathcal{U}_h(\bar{\Phi}'_\beta)$.

For $p\geq 1$, S^p denotes the space of real valued special semimartingales ($x = m + a$, $a_0=0$, a is previsible, of finite variation, m is a local martingale) equipped with the following norm

$$|x|_p = \Big\| [m,m]_1^{1/2} + \int_0^1 |da_s| \Big\|_{L^p(\Omega,\mathcal{F},P)}$$

(we indexe our processes either with $[0,1]$ or $\mathbb{R}_+$ for the sake of notational simplicity). S^0 denotes the space of all real valued semimartingales equipped with its usual metrizable but non locally convex Fréchet topology.

In the following we shall prove to the reader that the definition 1 is in fact much more practical than it seems at first glance :

Theorem 1 (Identification Theorem, cf. [16], [17])

a) There exists a one to one correspondence between the projective semimartingales and the linear, sequentially continous mappings from Φ into S° and any one of them is denoted by $S^\circ(\Phi')$.

b)There exists a one to one correspondence between S^p-projective semimartingales, the linear continuous mappings from Φ into S^p, the linear, sequentially continuous mappings from Φ into S^p and $\Phi'_\beta \tilde{\otimes} S^p$, i.e., the completed projective tensor product of S^p and Φ'_β , for $p \geqslant 1$ .This set will be denoted by $S^p(\Phi')$.

c)For any $X \in S^p(\Phi'), p \geqslant 1$, there exists an absolutely convex compact set $K \subset \Phi'$,such that, $\Phi'[K]$ is a separable Hilbert space and a $\Phi'[K]$ -valued S^p-semimartingale Y such that

$$\langle i_K(Y), \varphi \rangle = \tilde{X}(\varphi)$$

for any $\varphi \in \Phi$,where i_K denotes the canonical injection $\Phi'[K] \hookrightarrow \Phi'$.

Remarks

1)The parts (b) and (c) are the consequences of the nuclearity, the closed graph theorem and the fact that Φ is bornological.

2) The part (a) is more difficult to prove because S^0 is not locally convex .

3)In case (c) we say that X lives in $\Phi'[K]$.

A special case is the following :

Theorem 2 (c.f. [19])

Suppose that Φ'_β is metrizable .Then, for any $X \in S^\circ(\Phi')$, there exists a probability Q equivalent to $\mathbb{P}$ such that, under Q , X is in $S^1(\Phi')$ hence lives in some $\Phi'[K]$ as a (P or Q) semimartingale.

Remark

The theorem is valid even for the projective systems having

right and left limits. Let us also note that this result extends a recent result which says that in a nuclear, Fréchet space any probability measure has a Hilbert support.

II. Construction of the stochastic calculus

Let H be a bounded, previsible process with values in Φ. Then, there exists some $U \in \mathcal{U}_h(\Phi'_\beta)$ such that H is bounded and previsible in $\Phi[U^\circ]$. If $X \in S^\circ(\Phi')$, define

$$I(H)_t = \int_0^t \langle H_s, dX_s \rangle = \int_0^t (H_s, dX_s^U)$$

the last integral is a Hilbert space stochastic integral. Then I(H) is well defined and it is independent of the choice of U and this integral extends to the locally bounded processes (cf. [18]).

Integration by parts formula

If $X \in S^\circ(\Phi')$, $Z \in S^1(\Phi)$ then Z lives in some $\Phi[U^\circ]$, $U \in \mathcal{U}_h(\Phi'_\beta)$. Hence, from the Hilbert space integration (cf. [3]) we have

$$X(Z)_t = (X_t^U, Z_t)$$

$$= \int_0^t (X_{s-}^U, dZ_s) + \int_0^t (dX_s^U, Z_{s-}) + [\![X^U, Z]\!]_t$$

It is easy to show that the integrals are independent of the choice of U (cf. [18]), hence $[\![X^U, Z]\!]$ is also independent of the choice of U and is denoted by $[\![X, Z]\!]$. This result is called integration by parts formula and we write it as

$$\tilde{X}(Z_t)_t = \langle X_t, Z_t \rangle = \int_0^t \langle X_{s-}, dZ_s \rangle + \int_0^t \langle dX_s, Z_{s-} \rangle + [\![X, Z]\!]_t$$

Some applications of stochastic calculus

Let z be an $\mathbb{R}^d$-valued semimartingale with $z_0 = 0$. Then, for any $T \in \mathcal{D}'(\mathbb{R}^d)$, the following equation has a unique solution:

$$X_t = T - \int_0^t \partial_i X_{s-} \, dz_s^i + 1/2 \int_0^t \partial_{ij} X_{s-} \, d\langle z^{i,c}, z^{j,c} \rangle_s + $$
$$+ \sum_{0<s\leq t} [X_s - X_{s-} + \Sigma_i \, \partial_i X_{s-} \, . \Delta z_s^i]$$

In fact the solution is $T_* \delta_{z_t}$ and its uniqueness is straightforward with the use of the integration by parts formula.

For further applications we need the following regularity theorem which is a consequence of the Radon-Nikodym property of nuclear spaces :

Theorem 3 (Regularity Theorem, cf. [20])

Let X be a random field on $[0,1] \times \mathbb{R}^d$ such that $x \mapsto X(x)$ is C^∞ as an S^1-valued mapping. Then there exists an $\mathcal{E}(\mathbb{R}^d)$ -valued (i.e., the nuclear, Fréchet space of the infinitely differentiable functions on $\mathbb{R}^d$) semimartingale $\hat{X}$ such that, for any $x \in \mathbb{R}^d$, $X(x)$ and $\hat{X}(x)$ are undistinguishable.

As an application of this theorem we have

Corollary

Let $O \subset \mathbb{R}^d$ be open and $b: \mathbb{R}^d \times O \to \mathbb{R}^d$, $\sigma : \mathbb{R}^d \times O \mapsto \mathbb{R}^d \otimes \mathbb{R}^d$ be C^∞-vector fields with bounded derivatives. Then, there exists an $\mathcal{E}(O) \tilde{\otimes} \mathbb{R}^d$ -valued semimartingale π such that, for any $z \in O$, $\pi(z)$ is the solution of

$$dx_t(z) = b(x_t(z), z)\, dt + \sigma(x_t(z), z)\, dW_t$$

$$x_0(z) = \pi_0(z), \quad \pi_0 \in \mathcal{E}(O) \tilde{\otimes} \mathbb{R}^d,$$

where W is a d-dimensional Wiener process.

If z is a semimartingale with values in O (i.e., if δ_z is a semimartingale in $\mathcal{E}'(O)$: the distributions of compact support on O), then the integration by parts formula tells us that

$$\{ \pi_t(\omega, z_t(\omega)) = \langle \delta_{z_t(\omega)}, \pi_t(\omega, \cdot) \rangle ; (t,\omega) \in \mathbb{R}_+ \times \Omega \}$$

is a semimartingale and it can be developped as

$$\Pi_t(z_t) = \Pi_0(z_0) + \int_0^t b(\Pi_s(z_s), z_s)\,ds + \int_0^t \sigma(\Pi_s(z_s), z_s)\,dW_s +$$
$$+ \int_0^t (D_z\Pi_s)(z_{s-})\,dz_s + \frac{1}{2}\int_0^t (D_z^2\Pi_s)(z_{s-})\,d\langle z^c, z^c\rangle_s +$$
$$+ \int_0^t [D_z\sigma(\Pi_s(z_s), z_s) + D_x\sigma(\Pi_s(z_s), z_s)\, D_z\Pi_s(z_s)]\, d\langle W, z^c\rangle_s +$$
$$+ \sum_{0<s\le t}' [\Pi_s(z_s) - \Pi_s(z_{s-}) - D_z\Pi_s(z_{s-})\,\Delta z_s]$$

When the variable z is the initial condition of the diffusion and the semimartingale (z_t) is an Itô process then this formula is called Itô-Stratonovitch formula (cf.[12], [13], [14], [15], [19], [27]).In fact this result can be extended to the flows of the quasi-continuous semimartingales by the same method (cf.[2]) .

III.Some stochastic analysis

a) Continuity of the trajectories

In the definition of the projective processes we used the general nuclear spaces.In practice , all the nuclear spaces have some special properties which make the things easier. In this section we shall make a digression to study these cases.

Denote by R° the space of the (equivalence classes) of right continuous,measurable,real valued processes having left limits with the following metric :

$$d(x,o)=E(\inf(1,\sup_t |x_t|))\ .$$

Then we have the following

Theorem 4 (cf.[2])

Suppose that $J: \Phi \longrightarrow (R^\circ, d)$ is a continuous linear mapping and that Φ is separable.Then there exists a measurable process (X_t ; $t\in[0,1]$) with values in Φ' having almost surely strongly right continuous trajectories with left limits such that $\langle X_{\bullet}, \varphi \rangle$ belongs to the equivalence class of $J(\varphi)$ for any $\varphi \in \Phi$. Furthermore, there exists an increasing sequence of absolutely convex, compact subsets (K_n) of Φ' such that almost surely the process X lives in $\bigcup_{n\in\mathbb{N}} \Phi'[K_n]$.

Remarks

1) For $\Phi' = \mathcal{S}'$ this result has been proved in [5] (evidently without the second assertion of the theorem since $\mathcal{S}'$ itself is a countable union of a sequence of compact increasing sets), the extension to general Φ' and the fact that the trajectories are concentrated in a sequence of Hilbert spaces have been proved in [2]. Moreover [2] gives an Itô's formula for the semimartingales in nuclear spaces with the help of this result.

2) In practice all the nuclear that we use are separable, therefore the hypothesis of separability is not a restriction. To verify the continuity of J the easiest method is to show first its sequential continuity, then, if Φ is a Fréchet space or a countable inductive limit of Fréchet spaces (this is always the case in applications) this implies the continuity of J (cf. [1]).

3) We have fixed a metric on R° but this is not obligatory, any distance compatible with its vector space structure would be acceptable.

b) Applications to SPDE

In [26] we have applied the Theorem 4 to show the hypoellipticity of the stochastic partial differential operators :

We want to study the equation

(1) $du_t = -pu_t\, dt + q_i u_t\, dW_t^i + dh_t$

where $h \in S^\circ(\mathcal{E}(0))$, $0 \subset \mathbb{R}^d$ open, W is an $\mathbb{R}^n$-Wiener process, $p=p(t,\omega,x,\partial_x)$, $q_i=q_i(t,\omega,x,\partial_x)$, degree $p=2m$, degree $q_i=m_i \leq m$; and the coefficients of p and q_i are supposed to be C^∞ with uniformly bounded derivatives on the compact sets (they are local operators !) .

Definition

We say that the equation (1) is hypoelliptic if any solution u in $S^\circ(\mathcal{D}'(0))$ can be modified as an $\mathcal{E}(0)$-valued semimartingale given $h \in S^\circ(\mathcal{E}(0))$.

Let us denote by A_2 the space of the semimartingales k such that

$$k_t = k_0 + \int_0^t D_+ k_s \, ds + \int_0^t \partial_{W^i} k_s \, dW_s^i$$

with

$$E \int_0^t [|D_+ k_s|^2 + \sum_i |\partial_{W^i} k_s|^2] ds < +\infty \quad , \quad t \in \mathbb{R}_+.$$

Now we can announce :

Theorem 5 (Hypoellipticity Theorem; cf. [26])

Suppose that there exists some $s > 0$ such that for any $K \subset\subset 0$, for any bounded stopping time T, there exists $c, \bar{c} > 0$ depending only on K, s and T such that

$$\|u\|^2_{m+s-1,T} = E \int_0^T \|u_r\|^2_{m+s-1} \, dr \leq$$

$$\leq c(B_T(u,u) + \bar{c} \|u\|^2_{0,T})$$

for any $u \in A_2 \tilde{\otimes} \mathcal{D}_K(O)$, where $\|\cdot\|_\alpha$ denotes the Sobolev norm of order α , $\mathcal{D}_K(O)$ is the space of C^∞-functions on O whose supports are in K (compact) , B_T is the bilinear form defined by

$$B_T(u,v) = E \int_0^T ((D_+ + p_r)u_r, v_r)_0 \, dr + \frac{1}{2} E \sum_{i=1}^n \int_0^T (\partial_{W^i} u_r, \partial_{W^i} v_r) dr$$

and $(.,.)_0$ denotes the $L^2(0,dx)$-scalar product. Then the following stochastic partial differential equation is hypoelliptic :

$$du_t=(-p+((-1)^{m_i+1}/2)\ q_i^2)\ u_t\ dt + q_i u_t\ dW_t^i + dh_t .$$

Remarks

1) For the proof of this result we refer to [26] .

2)Note that these kind of equations have always been studied in the frame of Sobolev spaces and our approach ,as far as we know, is the first treatment of the equation in its natural setting,i.e. , as an equation in $S^\circ(\mathcal{D}'(\mathbb{R}^d))$.

c)Weak convergence of the processes

If $(X^n;n\in\mathbb{N})$ is a sequence of projective systems of continuous processes,then,we can define the weak covergence or convergence in law as the convergence of $(P_n^U;n\in\mathbb{N})$ on $\Omega^U = C([0,1],\Phi'(U))$, $U\in\mathcal{U}_h(\Phi'_\beta)$. We have in fact

Theorem 6 (cf.[6])

The sequence $(P_n;n\in\mathbb{N})$ of projective probabilities is tight if and only if ,for any $\varphi\in\Phi$, $(\tilde{X}_n(\varphi)_\circ;n\in\mathbb{N})$ is tight.

Suppose that Φ is a nuclear Fréchet space or inductive limit of a sequence of such spaces . Then (it is separable) as we have explained in (b),the projective systems of continuous or right continuous processes coincide with the strongly continuous or right continuous processes.Hence, for any $\{P^U ; U\in\mathcal{U}_h(\Phi'_\beta)\}$ on $\{C([0,1],\Phi'(U));U\in\mathcal{U}_h(\Phi'_\beta)\}$, there exists a (Radon) probability measure P on $C([0,1],\Phi'_\beta)$ such that $k'(U)(P)=P^U$ where $k'(U)$ is defined in an obvious way from $k(U)$. Let us announce a tightness result in this case :

Theorem 7

Suppose that Φ is a countable inductive limit of Fréchet spaces. Then a sequence of probabilities $(P_n; n \in \mathbb{N})$ on $C([0,1], \Phi'_\beta)$ is tight if and only if $(X^n_\cdot(\varphi); n \in \mathbb{N})$ is tight in $C([0,1], \mathbb{R})$ for any $\varphi \in \Phi$.

Remarks :

1) These results have not been found in the above order. In fact Theorem 7 was proved in [5] for the space of the tempered distributions, the Theorem 6 is proved in [6], moreover one can find there the case where Φ is a strict inductive limit of a sequence of Fréchet spaces. Relaxing the strictness hypothesis is then straightforward.

2) Of course all these results are true for the right continuous processes with left limits.

d) Additive processes

The characterization of additive processes on nuclear spaces is as easy as in the finite dimensional case. In the following we shall announce the results for the ordinary sense processes (not for the projective systems). Hence we suppose that Φ is separable and Souslin :

Theorem 8 (cf. [22])

Suppose that we are given a functional f_t on Φ for $t \in [0,T]$, $T>0$, of the following form :

$$f_t(\varphi) = \exp\left(i a_t(\varphi) - (1/2)\, \mathfrak{S}_t(\varphi,\varphi) + \int_{\Phi'} [\exp ix(\varphi) - 1 - ix(\varphi)]\, \mu_t(dx)\right)$$

such that

i) $t \mapsto a_t$ is continuous from $[0,T]$ into Φ'_β

ii) $t \mapsto \mathfrak{S}_t$ is continuous uniformly on the compact sets of $\Phi \times \Phi$, with values in the space of the bilinear forms on Φ, for any t, $\mathfrak{S}_t$ is nonnegative definite.

iii) $\mu_t, t \in [o,T]$, is an abstract set function on the measurable subsets of Φ' ,increasing with t continuously and there exists an absolutely convex compact $K \subset \Phi'$ such that $\mu_t(K^c \cap \cdot\,)$ is a Radon measure on Φ' for any $t \geq o$ with

$$\int_K p(x)^2 \ \mu_t(dx) < +\infty$$

for any continuous seminorm p on Φ'.

Then there exists an additive process X having right continuous trajectories with left limits in Φ'_β whose characteristic function is given by f_t. Conversely to any such additive process X, corresponds a functional f_t , $t \in [o,T]$,described as above .

Remarks

1)From Theorem 1 , X is a semimartingale if and only if the deterministic process $t \mapsto a_t(\varphi)$ is of finite variation for any $\varphi \in \Phi$.

2)Such an additive process can be decomposed as the sum of a deterministic process,a Gaussian process and a jump process as in the finite dimensional case (ie.,Paul Lévy's Theorem holds)(cf. [22]) .

A. S. Ustunel
2, Bd. A. Blanqui, 75013
Paris, France

REFERENCES

[1] Adasch,N.,Ernst,B.and Keim,D(1978).Topological Vector Spaces.Lect.N. in Math.Vol.639.Springer.

[2] Martias,C.(1983).Sur les supports des processus à valeurs dans des espaces nucléaires.Preprint.

[3] Métivier,M.&Pellaumail,J.(1980).Stochastic Integration. Academic Press.

[4] Mitoma,I.(1983).On the sample contunity of S'-processes. J.Math.Soc.Jap.Vol.35,no.4.

[5] Mitoma,I.(1984).Tightness of probabilities on C([o,1],S'). and on D([o,1],S').Annals of Probability .

[6] Fouque,J.P.(1984).La convergence en loi pour les processus à valeurs dans un espace nucléaire.A.I.H.P.,Vol.20,p.225.

[7] Nelson,E.(1967).Dynamical Theories of Brownian Motion. Princeton University Press.

[8] Pardoux,E.(1982).Equations du filtrage nonlinéaire de la prédiction et du lissage.Stochastics,Vol.6,p.193-231.

[9] Chaleyat-Maurel,M.&Michel,D.(1984).Hypoellipticity theorems and conditional laws.Z.f.W.

[10] Schaefer,H.H.(1971).Topological Vector Spaces.Grad.Texts in Math.Springer.

[11] Schwartz,L.(1973).Théorie des Distributions.3rd.Ed.Hermann Paris.

[12] Bismut,J-M.(1981).A generalized formula of Itô and other properties of stochastic flows.Z.f.W.,55,331-350.

[13] Bismut,J.M.(1981).Mécanique Aléatoire.Lect.N.in Math. Vol.866,Springer.

[14] Kunita,H.(1981).Some extensions of Itô's formula.Sém.de de Prob.XV,p.118-141.Springer.

[15] Rozovskii,B.L.(1973).A formula of Itô-Ventcell.Vestnik Moskou Univ. ,Ser.1 Math.Mch.26-32.

[16] Ustunel,A.S.(1982).A characterization of semimartingales on nuclear spaces.Z.F.W.,60,21-39.

[17] Ustunel,A.S.(1983).An erratum to "A characterization of semimartingales on nuclear spaces" after Laurent Schwartz.To appear in Z.f.W.

[18] Ustunel,A.S.(1982).Stochastic integration on nuclear spaces and its applications.A.I.H.P.,18,no.2,165-200.

[19] Ustunel,A.S.(1982).Some applications of stochastic integration in infinite dimensions.Stochastics,7,255-288.

[20] Ustunel,A.S.(1983).Analytic semimartingales and their boundary values.J.F.A.,Vol.51,no.2,p.142-158.

[21] Ustunel,A.S.(1983).Distributions-valued semimartingales and applications to control and filtering.Lect.N.in Cont.and Inf.Sci.Vol.61,p.314-325.Springer.

[22] Ustunel,A.S.(1984).Additive processes on nuclear spaces. The Annals of Prob.Vol.12,no.3,p.858-868.

[23] Ustunel,A.S.(1982).A generalization of Itô's formula. J.F.A.,47,143-152.

[24] Ustunel,A.S.(1983).Stochastic Feynman-Kac formula.J. d'Analy.Math.Vol.42,155-165.

[25] Ustunel,A.S.(1983).Some applications of stochastic calculus on nuclear spaces to the nonlinear stochastic problems.Nato ASI,Reidel Publ.Co.

[26] Ustunel,A.S.(1984).Hypoellipticity of the stochastic partial differential operators.Preprint.

[27] Ventzell,A.D.(1965).On equations of the theory of conditional processes.Theor.Prob.and Appl.10,357-361.

C. VAN DEN BROECK

STOCHASTIC LIMIT THEOREMS : SOME EXAMPLES FROM NONEQUILIBRIUM PHYSICS.

1. INTRODUCTION.

Fluctuations are the deviations of variables from their average value. Due to the discrete nature of the processes underlying the macroscopic phenomena, such fluctuations are present in every physico-chemical system. In many cases, it is found that these fluctuations are small. For instance, the dispersion of the number of gasparticles X(V) in a volume V is typically (Landau and Lifschitz, 1967)

$$<\delta X^2(V)> \approx <X(V)> \sim V \qquad (1)$$

We conclude that the amplitude of the fluctuations, $<\delta X^2(V)>^{1/2}$, is very small as compared to the average value $<X(V)>$ in a macroscopic system, i.e. for V large. Analogous results are found for the mean square displacement of a Brownian particle, along the x-axis, in function of time (D is the molecular diffusion coefficient) (Einstein, 1905)

$$<\delta x^2(t)> = 2Dt \sim t \qquad (2)$$

and for the light intensity I, scattered by N randomly distributed particles (E is the electrical field) (Berne and Pecora, 1976)

$$I = <E^2> \sim N \qquad (3)$$

From these results, one should not conclude that fluctuations are unimportant on the macroscopic scale. Firstly, some phenomena, whose effects can be observed macroscopically, are entirely due to the fluctuations, e.g. Brownian motion, light scattering, the escape of a thermally agitated particle out of a potential well or longitudinal dispersion (see section 4). Secondly, it is not always true that fluctuations are small. Large fluctuations can be associated to long range coherent behavior in a system close to a phase transition

L. Arnold and P. Kotelenez (eds.), Stochastic Space-Time Models and Limit Theorems, 179–189.

point (see sections 2 and 3). This particular behaviour is of interest to the physicist and mathematician alike. For the physicist, it is valuable to understand the mechanisms by which a system can undergo a transition from one macroscopic state to another one. For instance, it turns out that this mechanism is fundamentally different for a nonequilibrium transition towards a self-organized state and for a transition between different equilibrium structures, although the mathematical difficulties encountered are of the same type. These difficulties are due to the breakdown of the central limit theorem (in the sense that a probability density can not be approximated by a Gaussian probability density) : close to the transition point, the macroscopic quantities are no longer the result of a large number of "weakly correlated" microscopic contributions. Physicists have deviced a number of more or less successful techniques to calculate relevant features for these critical situations (see e.g. Ma, 1975). In doing so, they rely on physical argumentation rather than on rigorous mathematical results, which are rather scarce. In section 2, we briefly review the Van Kampen expansion of the master equation. In section 3, we discuss the breakdown of this expansion and propose an alternative "critical expansion". In section 4, we discuss a central limit theorem for the problem of longitudinal dispersion.

2. THE VAN KAMPEN EXPANSION OF THE MASTER EQUATION.

Let us consider a macroscopic system characterized by an extensive variable X, e.g. the total number of particles in a given total volume V. Such a variable is the sum of a large number of "microscopic contributions" $X_{\underset{\sim}{r}}$:

$$X = \sum_{\underset{\sim}{r}} X_{\underset{\sim}{r}} \tag{4}$$

i.e. $X_{\underset{\sim}{r}}$ is the number of particles in a small subvolume ΔV centered at the position vector $\underset{\sim}{r}$; the sum over $\underset{\sim}{r}$ runs over all the subvolumes constituting the total volume V. Generally speaking, the microscopic quantities $X_{\underset{\sim}{r}}$ will be "strongly fluctuating", i.e. $\langle \delta X_{\underset{\sim}{r}}^2 \rangle^{1/2}$ is of the same order of magnitude as $\langle X_{\underset{\sim}{r}} \rangle$ (e.g. the number of particles in a very small box will fluctuate between 0 and 1). On the other hand, we expect for the case of short range interaction forces between the particles that the spatial correlations are equally short ranged : $\langle \delta X_{\underset{\sim}{r}} \delta X_{\underset{\sim}{r}'} \rangle \simeq 0$ for $|\underset{\sim}{r} - \underset{\sim}{r}'| > l_c$, where l_c is the so-called correlation length. Hence, if one chooses subvolumes ΔV with linear dimension larger then l_c, one has

$$<\delta X_{\underset{\sim}{r}}\delta X_{\underset{\sim}{r}'}> \simeq <\delta X^2_{\underset{\sim}{r}}>\delta^{Kr}_{\underset{\sim}{r},\underset{\sim}{r}'}$$

If the system has translational invariance, then $<\delta X^2_{\underset{\sim}{r}}>$ is a constant and one concludes (see also eq. (1)) :

$$<\delta X^2> = \sum_{\underset{\sim}{r}}\sum_{\underset{\sim}{r}'} <\delta X_{\underset{\sim}{r}}\delta X_{\underset{\sim}{r}'}> = <\delta X^2_{\underset{\sim}{r}}> \frac{V}{\Delta V} \sim V \tag{5}$$

By the same token , one expects that the probability distribution, for X will be asymptotically, for large V, a Gaussian distribution.

Let us apply the above ideas to the case of an extensive time dependent random variable X(t). We expect that one can write :

$$X(t) = \overline{X}(t) + X(t) \tag{6}$$

where $\overline{X}(t)$ is the macroscopic part, proportional to the extensivity parameter V, and $\delta X(t)$ are small deviations of order $V^{1/2}$. This idea was exploited by Van Kampen (1961) to set up an asymptotic expansion for the probability distribution P(X,t) of a Markov process describing a physico-chemical system. For sake of simplicity, we consider the case of a Markov process X(t) taking values in $\mathbb{N}$. The conditional probability $P(X,t/X_0,t_0)$ obeys the following birth and death master equation :

$$\partial_t P(X,t/X_0,t_0) = \lambda(X-1)P(X-1,t/X_0,t_0)$$
$$+ \mu(X+1)P(X+1,t/X_0,t_0) - [\lambda(X)+\mu(X)]P(X,t/X_0,t_0) \tag{7}$$

The extensivity property is secured by the extensivity of the birth and death rates $\lambda \sim V$ and $\mu \sim V$. We are interested in the behavior of X(t) for large volume size $V \to \infty$. Van Kampen (1961) showed that, in this limit, the stochastic process $X(t)$ converges for finite times, to a deterministic process $\overline{X}(t) = \overline{x}(t)$. V, obeying the following ordinary differential equation

$$\frac{d\overline{x}(t)}{dt} = f(\overline{x}(t)) = \lim_{V\to\infty} \frac{\lambda(\overline{X}(t))-\mu(\overline{X}(t))}{V} \tag{8}$$

and that the scaled stochastic process :

$$u(t) = \frac{X(t) - \overline{X}(t)}{V^{1/2}} \tag{9}$$

converges, for finite times, to an Ornstein Uhlenbeck process obeying the following stochastic differential equation

$$du(t) = f'(\overline{x}(t)u(t) + Q^{1/2}dW(t) \tag{10}$$

f' stands for the derivative of f with respect to $\overline{x}$, $W(t)$ is the Wiener process and $Q = Q(\overline{x}(t))$ is the noise intensity, given by

$$Q(\overline{x}(t)) = \lim_{V\to\infty} \frac{\lambda(\overline{X}(t)) + \mu(\overline{X}(t))}{V} \tag{11}$$

For a rigorous formulation and proof of these theorems, see Kurz (1972). As far as the approach of the stationary state, i.e. the limit $t \to +\infty$, is concerned, one has to specify the stability properties of the macroscopic trajectory $\overline{x}(t)$. For instance, in the case of a bistable system, the steady state equation

$$f(\overline{x}^{st}) = 0 \tag{12}$$

possesses three solutions, two of which are asymptotically stable. In this case, the fluctuations corresponding to the transitions between these two states are of macroscopic order of magnitude, $<\delta X^2> \sim V^2$, and the convergence of the stochastic trajectory to the macroscopic one breaks down. A problem of particular interest is the onset of bistability at the so-called cusp bifurcation point. In this case, the unique stationary state is asymptotically but not linearly stable, i.e.

$$f(\overline{x}^{st}) = f'(\overline{x}^{st}) = f''(\overline{x}^{st}) = 0 \;;\; f'''(\overline{x}^{st}) < 0 \tag{13}$$

Since this situation lies at the borderline of the region of bistability ($<\delta X^2> \sim V^2$) and the "Gaussian regime" ($<\delta X^2> \sim V$, for all times, before bifurcation), we expect anomalous fluctuations, $<\delta X^2> \sim V^{2a}$ with $0.5 < a < 1$. These fluctuations can be studied by considering the scaled stochastic process

$$u'(t) = \frac{X(t) - \overline{X}(t)}{V^a} \tag{14}$$

The exponent a is choosen such that the resulting process u'(t) is nontrivial in the limit $V \to \infty$: if a is choosen too small, e.g. a = 1/2 as in eq. (9), one will find

$$\lim_{t \to +\infty} \langle u'^2(t) \rangle = +\infty$$

if a is taken too large, one obtains

$$\lim_{t \to \infty} \langle u'^2(t) \rangle = 0$$

Instead of pursuing these ideas dealing with "zero dimensional systems" (one scalar variable X), we will turn to the much more involved case of spatially distributed systems. Our motivation is twofold. Firstly, as it is clear from the arguments given at the beginning of this section, the breakdown of the result $\langle \delta X^2 \rangle \sim V$ must be associated to the arisal of long range spatial correlations. This is not surprising since we expect that the onset of new macroscopic states (which in more complicated cases exhibit spatial and/or temporal ordening) must involve long range coherence throughout the system. Secondly, the homogeneous description considered above is a mean field type of approximation since, in such a description the effect of inhomogeneous fluctuations is neglected. Such a description has been shown to lead to qualitatively wrong results in the theory of equilibrium phase transition (Ma, 1975).

3. CRITICALITY EXPANSION OF THE MULTIVARIATE MASTER EQUATION.

In order to take into account the effect of inhomogeneous fluctuations, it is necessary to formulate a master equation approach for spatially distributed system. We partition the system into an array of cubic subcells, centered at the position vectors $\{\underset{\sim}{r}\}$. The basic assumption is to suppose that the set of the particle numbers in these cells, $\{X_{\underset{\sim}{r}}\}$, constitute a Markov process. For the so-called reaction diffusion systems, two mechanisms are responsible for the time evolution of the multivariate probability distribution $P(\{X_{\underset{\sim}{r}}\},t)$: the chemical reactions, taking place inside each cell, being modelled by a birth and death process (cf. the previous section) and diffusion between adjacent cells, described by a random walk. As an example, we mention the following model (C. Van den Broeck et alii, 1980) :

$$X \xrightarrow{k_1} B \qquad A \xrightarrow{k_2} 2X \tag{15}$$

k_1 and k_2 are chemical rate constants and A is further used to denote the fixed number of particles, per cell, of a controlled chemical species. In terms of the generating function :

$$F(\{s_{\underset{\sim}{r}}\},t) = \sum_{\underset{\sim}{r}} \sum_{X_{\underset{\sim}{r}}=0}^{\infty} \prod_{\underset{\sim}{r}'} s_{\underset{\sim}{r}'}^{X_{\underset{\sim}{r}'}} P(\{X_{\underset{\sim}{r}'}\},t) \tag{16}$$

the multivariate master equation takes the following form :

$$\partial_t F(\{s_{\underset{\sim}{r}}\},t) = \sum_{\underset{\sim}{r}} [-k_1 (s_{\underset{\sim}{r}} - 1) \frac{\partial}{\partial s_{\underset{\sim}{r}}} \tag{17}$$

$$+ k_2 A (s_{\underset{\sim}{r}}-1)(s_{\underset{\sim}{r}}+1) + D \sum_{\underset{\sim}{l}} (s_{\underset{\sim}{r}+\underset{\sim}{l}} - s_{\underset{\sim}{r}}) \frac{\partial}{\partial s_{\underset{\sim}{r}}} \; F(\{s_{\underset{\sim}{r}}\},t)$$

The sum over $\underset{\sim}{l}$ runs over all the nearest neighbours and D, the jump rate, is related to Fick's diffusion coefficient D_F (which is a constant, independent of the cell size) by :

$$D = \frac{D_F}{(\Delta V)^{2/d}} \tag{18}$$

d is the dimensionality of the system.

Generally speaking, the solution of a multivariate master equation is very complicated, due to the large number of variables. Moreover, the cell size ΔV is not an obvious perturbation parameter, since it must be choosen small enough to describe the effect of inhomogeneous fluctuations. Rigorous results have been obtained by L. Arnold and P. Kotelenez (1983), but they are restricted to the case of linear reactions. For the more complicated model (15), we mention the instructive exact result for the equal time steady state pair correlation function ($d = 3$, $x_{\underset{\sim}{r}}(t) = X_{\underset{\sim}{r}}(t)/\Delta V$)

$$\lim_{\Delta V \to 0} \lim_{t\to\infty} = \frac{\langle \delta X_{\underset{\sim}{r}}(t)\delta X_{\underset{\sim}{r}'}(t)\rangle - \langle X_{\underset{\sim}{r}}(t)\rangle \delta^{Kr}_{\underset{\sim}{r},\underset{\sim}{r}'}}{\Delta V^2}$$

$$= \langle \delta x_{\underset{\sim}{r}} \delta x_{\underset{\sim}{r}'} \rangle^{st} - \langle x_{\underset{\sim}{r}} \rangle^{st} \delta(\underset{\sim}{r}-\underset{\sim}{r}') = \frac{\phi}{2\pi|\underset{\sim}{r}-\underset{\sim}{r}'|D_F} \exp(-|\underset{\sim}{r}-\underset{\sim}{r}'|/l_c) \tag{19}$$

$\phi = k_2 A$ is the material flux through the system. The correlation length l_c is given by :

$$l_c = (D_F/k_1)^{1/2} \tag{20}$$

We conclude that the correlation length is of the order of the distance, travelled by a particle between two reactive collisions. This result suggests that, for a more general reaction scheme, the correlation length will be of the form (20) with k_1 replaced by the total linearized decay rate $|f'(x^{st})|$ (cf. eq. 10), which we will further denote by δ :

$$l_c = (D_F/\delta)^{1/2} \sim \delta^{-\nu} \tag{21}$$

Note the divergence of the correlation length with a so-called classical critical exponent $\nu = 1/2$, at the approach of the bifurcation point $\delta = 0$. This divergence is responsible for the anomalous behavior of the global fluctuations

$$\lim_{V\to\infty} \frac{\langle \delta X^2 \rangle}{V} \sim \delta^{-\gamma} \qquad \text{with} \quad \gamma = 1 \text{ and } X = \int_V x_{\underset{\sim}{r}} d\underset{\sim}{r}$$

To check the validity of the result (21), a systematic expansion of the master equation was set up. The basic features of this criticality expansion are as follows :

- We anticipate the existence of long range spatial correlations. Small scale fluctuations are thus expected to be negligible and one can blow up the cell size ΔV. The number of cells is kept constant.
- By keeping the dominant terms in $\varepsilon = 1/\Delta V$ in the multivariate master equation, one shows that the discrete Markov process $X_{\underset{\sim}{r}}$, converges to a continuous Markov process $x_{\underset{\sim}{r}} . \Delta V$, where $x_{\underset{\sim}{r}}$ obeys the following Langevin equation with nonlinear drift but process independent noise

$$\frac{\partial x_{\underset{\sim}{r}}}{\partial t} = f(x_{\underset{\sim}{r}}) + D \sum_{\underset{\sim}{l}} (x_{\underset{\sim}{r}+\underset{\sim}{l}} - x_{\underset{\sim}{r}}) + \varepsilon^{1/2} F_{\underset{\sim}{r}}(t) \tag{22}$$

where $F_{\underset{\sim}{r}}$ is a multi-Gaussian white noise with correlation function :

$$<F_{\underset{\sim}{r}}(t)F_{\underset{\sim}{r}'}(t')> = \delta(t-t')\{Q(\overline{x}_{\underset{\sim}{r}})\delta^{Kr}_{\underset{\sim}{r},\underset{\sim}{r}'}$$

$$+ D\sum_{\underset{\sim}{l}}[(\overline{x}_{\underset{\sim}{r}+\underset{\sim}{l}}-\overline{x}_{\underset{\sim}{r}})\delta^{Kr}_{\underset{\sim}{r},\underset{\sim}{r}'} - (\delta^{Kr}_{\underset{\sim}{r}+\underset{\sim}{l},\underset{\sim}{r}'} - \delta^{Kr}_{\underset{\sim}{r},\underset{\sim}{r}'})(\overline{x}_{\underset{\sim}{r}}+\overline{x}_{\underset{\sim}{r}'})]\} \qquad (23)$$

The function f and Q are defined by the eqs. (8) and (11). We have thus established a central limit theorem, ensuring the convergence of a discrete Markov process to a continuous one. The particular status of this theorem is revealed when we calculate from it the stationary probability distribution. For the simplest example ($f(x) = -\delta x - x^3$, $Q = 1$, this corresponds to the Schögl model), one finds :

$$P_{st}(\{x_{\underset{\sim}{r}}\}) = \exp\{\frac{2\varepsilon}{Q_{st}}$$

$$\sum_{\underset{\sim}{r}}[\frac{\delta}{2}(x_{\underset{\sim}{r}}-\overline{x}_{st})^2 + \frac{1}{4}(x_{\underset{\sim}{r}}-\overline{x}_{st})^4 + \frac{D}{4}\sum_{\underset{\sim}{l}}(x_{\underset{\sim}{r}+\underset{\sim}{l}}-x_{\underset{\sim}{r}})^2]\} \qquad (24)$$

This probability does not belong to the class of infinitely divisible laws, even in the homogeneous case ($D = 0$), due to the presence of a quartic term in the exponential. This reflects the coherence associated with bifurcation : the system can no longer be partitioned into a collection of weakly correlated subsystems. On the other hand the exponential in (24) displays the familar Landau Ginzburg type of potential in discrete (position) space. This establishes the connection between the Master Equation approach and the theory of equilibrium critical phenomena. Relevant properties, such as the spatial correlation, critical dimensionality and critical exponents can be obtained from renormalization group theory (Ma, 1975). In particular, the correlation length is divergent at the approach of the bifurcation point for dimensions $d > 1$, validating our perturbative expansion in these cases. For $d = 1$, the system exhibits no critical behaviour at the crossing of the bifurcation point. The divergence of the correlation on length l_c in $d > 1$, is not always characterized by the classical critical exponent $\nu = 1/2$. For instance, for the case of a probability distribution of the type (24), one has $\nu = 1$ and $\gamma = 1.75$ for $d = 2$ and $\nu \approx 0.64$ and $\gamma \approx 1.25$ in $d = 3$. Only for a dimension larger then the so-called critical dimensionality $d_c = 4$, one recovers the classical critical exponents $\nu = 1/2$ and $\gamma = 1$ (Ma, 1975).

4. A CENTRAL LIMIT THEOREM FOR LONGITUDINAL DISPERSION.

In the previous sections, we discussed the stochastic properties of an extensive variable X in the limit of a large volume size V. In this section, we will illustrate how analogous results are obtained for the case of layered systems in the long time limit.

We consider a system, consisting of n parallel horizontal layers i = 1,..., n. In each of these layers, a fluid is flowing with a velocity u_i, along the horizontal x-axis. We follow the stochastic trajectory of a suspended particle, which performs a random walk in the vertical direction over the n layers, while being carried along with the local fluid flow, in each layer. The transition probabilities, per unit time, from layer i to i ± 1 are $k_i^{\pm}$. For simplicity of notation, we set $k_{-1}^{+} = k_{n+1}^{-} = 0$, while $k_1^{-} = k_n^{+} = 0$. The probability densities P(x,i,t) to find the particle at the position x at time t in the layer i satisfy the following set of conservation equations :

$$\frac{\partial}{\partial t} P(x,i,t) = k_i^{+} P(x,i-1,t) + k_i^{-} P(x,i+1),t)$$
$$- (k_i^{+} + k_i^{-})P(x,i,t) - \frac{\partial}{\partial x}[u_i P(x,i,t)] \qquad (25)$$

Before proceeding, we note that this model has applications in problems such diverse as petro-engineering, chromatography, Taylor diffusion, spin diffusion in one dimensional chains, residence time problems in reactors, phase diffusion in limit cycles, etc. (see C. Van den Broeck and R.M. Mazo (1984)).

In analogy with eq. (4), the displacement of the suspended particle at time t, x(t), is the sum of a large number of displacements Δx_i during subsequent time intervals $[(i-1)t, i\Delta t]$, i = 1,...,N ($N\Delta t = t$). Moreover, as the particle "wanders" up and down the layers, it forgets about its earlier positions. Hence, if Δt is choosen larger than the characteristic memory time, then the displacements Δx_i are (almost) independent one of another and we conclude (cf. eq. (5)) :

$$<\delta x^2(t)> \sim t \qquad (26)$$

The proportionality constant, called the effective longitudinal diffusion coefficient K has been calculated exactly for the above model (C. Van den Broeck and R.M. Mazo (1983) :

$$\lim_{t\to+\infty} \frac{<\delta x^2(t)>}{2t} = K = \sum_{r=1}^{n-1} \frac{\left[\sum_{i=1}^{n} \sum_{j=1}^{r} (u_i - u_j)\, P_i^{st} P_j^{st}\right]^2}{k_r^+ \, P_r^{st}} \qquad (27)$$

with

$$P_r^{st} = N\, k_1^+ k_2^+ \ldots \kappa_{r-1}^+ k_{r+1}^- \ldots k_{n-1}^- k_n^- \qquad (28)$$

and N is a normalisation constant

$$\left(\sum_{r=1}^{n} P_r^{st} = 1 \right)$$

It must be stressed that the result (26) heavily rests upon the sufficiently fast decay of the time correlation function. This is best illustrated by considering the continuum limit of the above model in the case of a linear velocity field. The vertical direction (perpendicular to the layers) will be denoted by y, and the velocity field reads $u(y) = \alpha y$. The probability density $P(x,y,t)$ obeys the following Fokker Planck equation

$$\frac{\partial}{\partial t} P(x,y,t) = D \frac{\partial^2}{\partial y^2} P(x,y,t) - \frac{\partial}{\partial x} \left[u(y) P(x,y,t) \right] \qquad (29)$$

A straightforward calculation yields :

$$<\delta x^2(t)> = \frac{2}{3} \alpha^2 D t^3 + <\delta x^2(t=0)> \qquad (30)$$

hence

$$\lim_{t\to+\infty} \frac{<\delta x^2(t)>}{t} = +\infty$$

The reason for the breakdown of the result (26) is the existence of an infinite correlation time. A particle never forgets its initial position, as it is shown by the exact result for the time correlation function :

$$\langle \delta x(t)\delta x(t')\rangle = \langle \delta x^2 (\min(t,t')\rangle \qquad (31)$$

C. Van den Broeck
Dept. Natuurkunde
Vrije Universiteit Brussel
Pleinlaan 2, 1050 Brussel, Belgium.

REFERENCES

B.J. Berne and R. Pecora (Eds.), *Dynamic Light Scattering* (Wiley, New York, 1976).

A. Einstein, Ann. Phys. **17**, 549 (1905).

P. Kotelenez, Ph. Thesis, University of Bremen (1983).

T.G. Kurz, J. Chem. Phys. **57**, 2976 (1972).

S.K. Ma, *Modern Theory of Critical Phenomena* (Benjamin, N.Y., 1975).

M. Malek Mansour, C. Van den Broeck, G. Nicolis and J.W. Turner, Ann. Phys. **131**, 283 (1981).

C. Van den Broeck, M. Malek Mansour and J. Houard, Physica **101A**, 167 (1980).

C. Van den Broeck and R.M. Mazo, Phys. Rev. Lett. **51**, 1309 (1983).

C. Van den Broeck and R.M. Mazo, J. Chem. Phys. (to appear, 1984).

N.G. Van Kampen, Can. J. Phys. **39**, 551 (1961).

B. Grigelionis, R. Mikulevičius

ON THE FUNCTIONAL LIMIT THEOREMS

Introduction

In the recent years an increasing interest is paid to the functional limit theorems for infinite dimensional stochastic processes, related to the investigations of stochastic space-time models in physics, chemistry, biology and other fields as well as in the general theory of stochastic evolution equations (see, e.g. [1] - [20]). Most of the known results are concerned with the cases of random processes taking values in Hilbert spaces, spaces of measures and nuclear spaces, in particular, spaces of distributions. Note that sometimes it is necessary to consider a weak convergence of probability measures on nonmetrisable topological spaces. For example, in the case of weak convergence to the solutions of stochastic evolution equations there are considered semimartingales $X_t = X_0 + A_t + Z_t$, $t \geq 0$, taking values in the rigged Hilbert space $V \subset H \equiv H' \subset V'$, where V is a separable reflexive Banach space, which is dense in H, moreover the imbeddings $V \to H$ and $H' \to V'$ are continuous, A_t, $t \geq 0$, is a V'-valued stochastic process with the finite variation on each finite time interval, Z_t, $t \geq 0$, is a H-valued

L. Arnold and P. Kotelenez (eds.), Stochastic Space-Time Models and Limit Theorems, 191–215.

semimartingale, satisfying the condition $\int_{0}^{t} |X_s|_V |dA_s|_{V'} < \infty$ for each $t > 0$. Under some natural assumptions it reduces to the investigation of stochastic processes with the paths in the function space $X = D_{[0,\infty)}(H) \cap \cap L^p_{loc}(V,N)$, where $D_{[0,\infty)}(H)$ is the Skorokhod space,

$$L^p_{loc}(V,N) = \{\hat{w}(\cdot): \int_0^t |\hat{w}(s)|^p_V dN_s < \infty, \forall t \geq 0\},$$

$p \geq 1$, and N is an increasing continuous function. The space X is considered to be equipped with the supremum of J_1-topology and the topology of weak convergence in $L^p_{loc}(V,N)$. It is important to note, that such topological spaces are completely regular and the criterion of Prokhorov for weak relative compactness of the sets of probability measures on them is still valid (see [21]).

It is well known that in the functional limit theorems for finite dimensional as well as infinite dimensional stochastic processes the method based on the martingale characterization of limiting points, combined with the known criteria of relative compactness and the laws of large numbers, is rather effective and universal. Following [22], the aim of this paper is the formalization of this method for arbitrary topological spaces and its applications to the weak convergence of semimartingales taking values in the rigged Hilbert spaces.

1. Martingale characterization and weak convergence of probability measures on topological spaces

Let X be an arbitrary topological space, $\mathcal{B}(X)$ be the σ-algebra of the Borel subsets, $B = \{\mathcal{B}_t, t\geq 0\}$ be some increasing family of sub-σ-algebras, $\mathcal{B}_t \subset \mathcal{B}(X)$, $t \geq 0$, $B_+ = \{\mathcal{B}_{t+}, t\geq 0\}$, $\mathcal{B}_{t+} = \bigcap_{\varepsilon>0} \mathcal{B}_{t+\varepsilon}$, $t \geq 0$, $\hat{P}$ be a probability measure on $\mathcal{B}(X)$, $\{M_t^i(x), t\geq 0, x\in X, i\in I\}$ a family of functionals being right continuous in t with left limits, where I is some set. We shall assume that the measure $\hat{P}$ is uniquelly defined by the assumptions that $\hat{P}|\mathcal{B}_0 = \mu_0$ is given and for each $i \in I$ M^i are $(B_+,\hat{P})$-local martingales (for terminology see, e.g., [23] - [25]).

Let us consider further a sequence $\{X_n, n\geq 1\}$ of random elements, defined on probability spaces $(\Omega_n, \mathcal{F}_n, P_n)$ with the given filtrations of σ-algebras $F_n = \{\mathcal{F}_{nt}, t\geq 0\}$ and taking values in X, $n \geq 1$. Denote $\hat{P}_n = P_n \circ X_n^{-1}$, $\hat{P}_n|\mathcal{B}_0 = \mu_0^{(n)}$, $n \geq 1$. We shall assume that we are given the families $\{M^{(n),i}, i\in I\}$ of (F_n, P_n)-local martingales, in terms of which the conditions of weak convergence $\hat{P}_n \Rightarrow \hat{P}$ as $n \to \infty$ will be formulated.

Denote by $\mathcal{T}(F_n)$ the class of stopping times (s.t.) with respect to the filtration F_n, and for each $T > 0$ by $\Sigma(T)$ the class of such sequences $\{S_n, n\geq 1\}$ of s.t. $S_n = S_n(T) \in \mathcal{T}(F_n)$, that $S_n \leq T$ and

$$\lim_{n\to\infty} P_n\{S_n<T\} = 0 . \qquad (1)$$

Let M_X be a class of limiting points of the sequence $\{\hat{P}_n, n\geq 1\}$ and K be a class of increasing functions $G: R_+ \to R_+$ such that $G(t)/t \to \infty$ as $t \to \infty$.

Assume that for each $\nu \in M_X$ and $i \in I$ there exists a denumerable dense subset $Q = Q^{\nu,i} \subset R_+$ and $G = G^{\nu,i} \in K$, such that the following conditions are satisfied:

(α) $\mathcal{B}_t = \sigma\{f: f\in B_t\}$ for each $t \in Q$, where B_t is some set of ν - a.e. continuous bounded functions on X;

(β) for each $t \in Q$ and $i \in I$ the functional $M_t^i(\cdot)$ is ν - a.e. continuous;

(γ) for some sequence $\{T_k, k\geq 1\}$, $0 \leq T_k \uparrow \infty$ there exist sequences $\{S_n^{(k)}, n\geq 1\} \in \Sigma(T_k)$, such that for each $t \in Q$, $i \in I$ and $k \geq 1$

$$\sup_{n\geq 1} E_n\, G(|M^{(n),i}_{t\wedge S_n^{(k)}}|) < \infty \qquad (2)$$

The following assertion holds.

Theorem 1. Assume that:

1) the sequence $\{\hat{P}_n, n\geq 1\}$ is weakly relative compact;

2) for some subclass of sets $B \in \mathcal{B}_0$, defining μ_0,

$$\lim_{n\to\infty} \mu_0^{(n)}(B) = \mu_0(B);$$

3) the assumptions (α) - (γ) are satisfied;

4) for each $\varepsilon > 0$, $t \in \bigcup_{\nu\in M_X} Q^{i,\nu}$, $i \in I$

$$\lim_{n\to\infty} P_n\{|M_t^{(n),i}-M_t^i(X_n)| > \varepsilon\} = 0 .$$

Then $\hat{P}_n \Rightarrow \hat{P}$ as $n \to \infty$.

P r o o f of the theorem 1 follows directly from the results of § 1 of the paper [22].

2. Canonical form of semimartingales with values in rigged Hilbert spaces

Stochastic processes, considered in this paragraph, are assumed to be defined on a probability space $(\Omega,\mathcal{F},P)$ and adapted to a filtration of σ-algebras $F = \{\mathcal{F}_t, t\geq 0\}$.

For a normed vector space E we shall denote by $\mathcal{V}(E)$ a class of right continuous E-valued processes with left limits having P - a.e. finite variation on each finite time interval, $\mathcal{A}(E)$ a subclass of processes of $\mathcal{V}(E)$ which have summable variation, $\mathcal{M}(E)$ a class of uniformly integrable E-valued (P,F)-martingales, $\mathcal{M}^2(E)$ a subclass of $\mathcal{M}(E)$ of square integrable martingales, $\mathcal{S}(E)$ a class of E-valued (P,F)-semimartingales, $\mathcal{S}_p(E)$ a class of E-valued (P,F)-special semimartingales. As usual for each class of processes C we denote by C^c a subclass of continuous elements of C and by C_{loc} a class of processes $C = \{C_t, t\geq 0\}$, for which there exists a sequence $\{T_n, n\geq 1\}$, $T_n \in \mathcal{T}(F_n)$, $T_n \uparrow \infty$, such that $\{C_{t\wedge T_n}-C_0, t\geq 0\} \in \mathbf{C}$ for all $n \geq 1$. $\mathcal{P}$ will denote a σ-algebra of F-predictable subsets of $[0,\infty) \times \Omega$. For two normed vector spaces $E_1 \supseteq E_2$ and $e \in E_1$ we shall denote by $|e|_{E_2}$

the norm of e in E_2, if $e \in E_2$, and $= \infty$ if $e \notin E_2$.

Let us consider a rigged Hilbert space $V \subset H \equiv H' \subset V'$ and a stochastic process $X = \{X_t = A_t + Z_t, t \geq 0\}$, where $A \in \mathcal{V}(V')$, $A_0 = 0$, $Z \in \mathcal{S}(H)$ and for each $t > 0$ P a.e. $\int_0^t |X_s|_V d|A|_s < \infty$; $|A|_t = \int_0^t |dA_s|_{V'}$. Denote by $p(du,dx)_\Pi$ the jump measure of X, $\Pi(du,dx)$ the dual predictable projection (the compensator) of $p(du,dx)$, $q(du,dx) = p(du,dx) - \Pi(du,dx)$.

Let $h: H \to H$ be a continuous bounded function, $h(x) = x$ for $|x|_H \leq \frac{1}{2}$ and $= 0$ for $|x|_H$ 1.

Theorem 2. The paths of the process X P - a.e. are H-valued right continuous with left limits in the strong topology of H. Moreover

$$|X_t|_H^2 = |X_0|_H^2 + 2\int_0^t X_s dA_s + 2\int_0^t A_{s-} dZ_s + $$
$$+ |Z_t|_H^2 - \sum_{s \leq t} |\Delta A_s|_H^2, \ t \geq 0, \tag{4}$$

and the following unique decompostion holds:

$$X_t = X_0 + \alpha_t^h + X_t^c + \int_0^t \int_{H\setminus\{0\}} h(x) q(du,dx) + \int_0^t \int_{H\setminus\{0\}} (x - h(x)) p(du,dx), \tag{5}$$

where $\alpha^h \in \mathcal{A}_{loc}(V') \cap \mathcal{P}$, $X^c \in \mathcal{M}_{loc}^c(H)$, $\Delta\alpha_t^h = t \geq 0$, $= \int_{H\setminus\{0\}} h(x)\ \Pi(\{t\} \times dx)$, $t \geq 0$, and there exist $C \in \mathcal{V}(V')$, $S \in \mathcal{S}(H)$, such that P - a.e. for each $t > 0$ $\int_0^t |X_s|_V\, d|C|_s < \infty$ and $\alpha^h = C + S$.

P r o o f . The first part of theorem 2 is proved mainly following the papers [26], [27], containing more special results.

Let $Z = D + M$, $D \in V(H)$, $M \in M_{loc}(H)$,

$$\beta_t = \inf\{s: |A|_s > t\}, \ |D|_t = \int_0^t |dD_s|_H, \ t \geq 0.$$

Without restriction of generality we shall assume, that

$$\beta_0=0, |A|_\infty+|D|_\infty+|X_0|_H^2+\int_0^\infty |X_s|_V d|A|_s+\sup_{t\geq 0}|M_t|_H\leq 1.$$

We shall need later several lemmas. By means of the standard change of time we obtain the following assertion.

Lemma 1. If g is $F \otimes B(\mathbb{R}^1)$-measurable function and $\int_0^t |g(u)|d|A|_u < \infty$, then

$$\int_0^t g(u)d|A|_u = \int_0^{|A|_t} g(\beta_s)ds .$$

In addition, $|A|_{\beta_t-} - |A|_{\beta_s} \leq t - s, \ s < t$.

Denote

$$\psi_n^{(1)}(s) = \frac{i}{2^n}, \ s \in [\frac{i}{2^n},\frac{i+1}{2^n}), \ i = 0, \pm 1, \ldots,$$

$$\psi_n^{(2)}(s) = \frac{i+1}{2^n}, \ s \in (\frac{i}{2^n},\frac{i+1}{2^n}], \ i = 0, \pm 1,\ldots$$

Lemma 2. Let $\{g_k, k\geq 1\}$ be a sequence of V-valued $B \otimes B(\mathbb{R}^1)$-measurable functions such that $g_k(s)=0$ for $s\notin[0,1]$ and $E[\int_0^1 |g_k(s)|_V ds]<\infty$, $k\geq 1$. Then there exists a sequence $n_j \uparrow \infty$ such that for almost all $u \in [0,1]$ with respect to the Lebesgue measure

$$\lim_{j\to\infty} E[\int_0^1 |g_k(s)-g_k(\psi_{n_j}^{(i)}(s-u)+u)|_V ds] = 0, \ k\geq 1, \ i=1,2.$$

P r o o f . For each $k \geq 1$ and $\varepsilon > 0$ there exists a $\mathcal{F} \otimes \mathcal{B}(\mathbb{R}^1)$-measurable function $g_{k,\varepsilon}(s)$, continuous in s, such that

$$E[\int_{-\infty}^{\infty} |g_k(s) - g_{k,\varepsilon}(s)|_V ds] \leq \varepsilon$$

It follows that for each $t \in \mathbb{R}^1$, $k \geq 1$

$$\lim_{h\to 0} [\int_{\infty}^{\infty} |g_k(s+t+h)-g_k(s+t)|_V ds] = 0 .$$

and therefore as $n \to \infty$

$$\eta_{k,n}^{(i)}(t) = E[\int_0^{\infty} |g_k(s+\psi_n^{(i)}(t)) - g_k(s+t)|_V ds] \to 0 ,$$

moreover $\eta_{k,n}^{(i)}(t) = 0$ for $t > 2$.

Thus as $n \to \infty$

$$\int_0^{\infty} \int_{-1}^{\infty} E|g_k(s+\psi_n^{(i)}(t)) - g_k(s+t)|_V ds\, dt =$$

$$= \int_0^{\infty} \int_{t-1}^{\infty} E|g_k(s+\psi_n^{(i)}(s-t)) - g_k(s)|_V ds\, dt \to 0$$

and we can find $n_k \uparrow \infty$ such that

$$\sum_{i=1}^{2} \sum_{j=1}^{k} \int_0^{\infty} \int_{t-1}^{\infty} E|g_j(t+\psi_n^{(i)}(s-t))-g_j(s)|_V ds\, dt \leq 2^{-k}$$

for each $n \geq n_k$. So we have that for almost all $t \in [0,1]$ and $k \geq 1$

$$\sum_{i=1}^{2} \sum_{j=1}^{\infty} \int_0^{\infty} E|g_k(t+\psi_{n_j}^{(i)}(s-t)) - g_k(s)|_V ds < \infty .$$

Lemma 2 is proved.

Let $\{e_j, j\geq 1\}$ be a basis of H consisting elements from V, $Z_t^{(k)} = \sum_{j=1}^{k} (Z_t, e_j)_H e_j$. Further we agree for vector-valued function f, defined on $[0,\infty)$, to assume that $f(\infty) = 0$.

Using lemma 1 we have that for each $k \geq 1$

$$E[\int_0^1 (|X_{\beta_s}|_V + |Z_{\beta_s}^{(k)}|_V) ds < \infty .$$

Thus from lemma 2 we deduce the following assertion.

Corollary 1. There exists a sequence of the imbedded partitions $0 = s_0^{(n)} < s_1^{(n)} < \ldots < s_{l_n}^{n} = 1$, $\max_{1\leq j\leq l_n} (s_{j+1}^{(n)} - s_j^{(n)}) \to 0$, as $n \to \infty$, such that for all $k \geq 1$, $i = 1,2$

$$\lim_{n\to\infty} E[\int^{\infty} |X_{\beta_s} - \overline{X}_{\beta_{\varphi_n^{(i)}(s)}}|_V \, ds] = 0$$

and

$$\lim_{n\to\infty} E[\int_0^{\infty} |Z_{\beta_s}^{(k)} - \overline{Z}_{\beta_{\varphi_n^{(i)}(s)}}^{(k)}|_V \, ds] = 0 ,$$

where

$$\varphi_n^{(1)}(s) = s_j^{(n)}, \; s \in [s_j^{(n)}, s_{j+1}^{(n)}), \; j = 1,\ldots,l_n ,$$

$$\varphi_n^{(2)}(s) = s_{j+1}^{(n)}, \; s \in (s_j^{(n)}, s_{j+1}^{(n)}], \; j = 1,\ldots,l_n ,$$

$$\overline{X}_0 = \overline{Z}_0^{(k)} = 0, \; \overline{X}_s = X_s, \; \overline{Z}_s^{(k)} = Z_s^{(k)} \quad \text{for} \quad s > 0.$$

Let further $\tau_j^{(n)} = \beta_{s_j^{(n)}}$, $j = 1,\ldots,l_n$, $X_n^{(1)}(s) = 0$ for $0 \leq s < \tau_1^{(n)}$, $= X_{\tau_j^{(n)}}$, for $s \in [\tau_j^{(n)}, \tau_{j+1}^{(n)})$, $j = 1,\ldots,l_n$, $X_n^{(2)}(s) = X_{\tau_1^{(n)}}$ for $0 < s \leq \tau_1^{(n)}$,

$= X_{\tau_{j+1}^{(n)}}$ for $s \in (\tau_j^{(n)}, \tau_{j+1}^{(n)}]$, $j = 1,\dots,l_n-1$.

The processes $Z_n^{(k),i}$, $i = 1,2$, are defined in the analogous way.

Corollary 2. The following equalities hold:

$$\lim_{n\to\infty} E[\int_0^\infty |X_u - X_n^{(i)}(u)|_V d|A|_u] = 0,\ i = 1,2,$$

$$\lim_{n\to\infty} E[\int_0^\infty |Z_u^{(k)} - Z^{(k),i}(u)|_V d|A|_u] = 0,\ i = 1,2,\ k\geq 1.$$

Indeed, from lemma 1 and corollary 1 it follows that

$$\lim_{n\to\infty} E[\int_0^\infty |X_u - X_n^{(i)}(u)|_V d|A|_u] = \lim_{n\to\infty} E[\int_0^\infty |X_{\beta_s} - X_n^{(i)}(\beta_s)|_V ds] \leq \lim_{n\to\infty} \sum_{i=1}^2 E[\int_0^\infty |X_{\beta_s} - \overline{X}_{\beta_{\varphi_n^{(i)}(s)}}|_V ds] = 0$$

The second equality is proved similarly.

Corollary 3. We have, that

$$E[\sup_{u,n} |X_n^{(i)}(u)|_V] < \infty,\ i = 1,2.$$

It follows from the corollary 2 and the step form of $X_n^{(i)}$. It is easy to check that

$$|X_{\tau_{k+1}^{(n)}}|_H^2 = |X_0|_H^2 + 2\int_0^{\tau_{k+1}^{(n)}} X_n^{(2)}(s)dA_s + 2\sum_{j=0}^k X_{\tau_j^{(n)}}(Z_{\tau_{j+1}^{(n)}} - Z_{\tau_j^{(n)}}) + \sum_{j=0}^k (|Z_{\tau_{j+1}^{(n)}} - Z_{\tau_j^{(n)}}|_H^2 - |A_{\tau_{j+1}^{(n)}} - A_{\tau_j^{(n)}}|_H^2) =$$

$$|X_0|_H^2 + 2\int_0^{\tau_{k+1}^{(n)}} X_n^{(2)}(s)dA_s + 2\sum_{j=0}^k (X_{\tau_j^{(n)}} - Z_{\tau_j^{(n)}})(Z_{\tau_{j+1}^{(n)}} -$$

$$- Z_{\tau_j^{(n)}}) + |Z_{\tau_{k+1}^{(n)}}|_H^2 - \sum_{j=0}^{k} |A_{\tau_{j+1}^{(n)}} - A_{\tau_j^{(n)}}|_H^2 . \tag{6}$$

Using the inequality $2ab \le \varepsilon^2 a^2 + \frac{b^2}{\varepsilon^2}$, $\varepsilon > 0$, from (6) and corollary 2 we have that

$$E[\sup_{n,k} |X_{\tau_k^{(n)}}|_H^2] < \infty \tag{7}$$

and then

$$E[\sup_{n,k} |A_{\tau_k^{(n)}}|_H^2] < \infty . \tag{8}$$

Lemma 3. P - a.e. for each $t > 0$ $A_t \in H$, A is right continuous with left limits in the weak topology of H and

$$E[\sup_{t\ge 0}(|A_t|_H^2 + |A_{t-}|_H^2)] < \infty \tag{9}$$

P r o o f . Set $I = I(w) = \bigcup_{n,k} \{\tau_k^{(n)}(w)\} \cap [0,\infty)$. Remark that the closure $\overline{I}$ is the support of the measure $d|A|_t$ on $[0,\infty)$. So far as according to (8) P - a.e. $\sup_{t\in I} |A_t|_H^2 < \infty$, the set $\{A_t, t\in I\}$ is a weakly compact subset of H. If $t_n \in I$, $t_n > t$ and $t_n \downarrow t$ (respectively, $t_n<t$, $t_n \uparrow t$), then $A_{t_n} \to A_t$ (respectively, $A_{t_n} \to A_{t-}$) and $|A_t|_H \le \sup_n |A_{t_n}|_H$ (respectively, $|A_{t-}|_H \le \sup_n |A_{t_n}|_H$). It is clear that $\overline{I}^c = [0,\infty)\setminus\overline{I}$ is the union of a finite or denumerable number of open intervals and A is constant on each of them. Lemma 3 is proved.

Lemma 4. Let $\{a_n, n\ge 1\}$ be a sequence of H-valued predictable processes such that $P\{\sup_{n,t}|a_n(t)|_H<\infty\}=1$ and for each $t \ge 0$ P - a.e. $\lim_{n\to\infty} a_n(t) = 0$ in the weak topology of H. Then for each $\varepsilon > 0$

$$\lim_n P\{\sup_t |\int_0^t a_n(s)dZ_s| > \varepsilon\} = 0 .$$

P r o o f . Denote Π^k the projector to the subspace, generated by the k first elements of a basis in H. Then

$$P\{\sup_t |\int_0^t a_n(s)dZ_s| > \varepsilon\} \le P\{\sup_t |\int_0^t a_n(s)dD_s| > \frac{\varepsilon}{2}\} +$$

$$+ P\{\sup_t |\int_0^t a_n(s)dM_s| > \frac{\varepsilon}{2}\} \tag{10}$$

and for each $k \ge 1$ as $n \to \infty$

$$P\{\sup_t |\int_0^t a_n(s)dD_s| > \frac{\varepsilon}{2}\} \le P\{\sup_t |\int_0^t \Pi^k a_n(s)dD_s| > \frac{\varepsilon}{4}\} +$$

$$+ P\{[\sup_t |a_n(t)|_H |(I-\Pi^k)D|_\infty] > \frac{\varepsilon}{2}\} \to 0 \tag{11}$$

according to the assumptions of the lemma and the fact that $|I-\Pi^k)B|_\infty \to 0$ in probability as $k \to \infty$.

Using the known inequality (see [28]) for each $\delta > 0$ and $\varepsilon > 0$

$$P\{\sup_t |\int_0^t a_n(s)dM_s| > \frac{\varepsilon}{2}\} \le P\{\sup_t |\int_0^t \Pi^k a_n(s)dM_s| > \frac{\varepsilon}{4}\} +$$

$$+ P\{\sup_t |\int_0^t a_n(s)d[(I-\Pi^k)M_s]| > \frac{\varepsilon}{4}\} \le$$

$$\le P\{\int^\infty |\Pi^k a_n(s)|_H^2 \, d\langle M\rangle_s > \delta\} + \frac{16\delta}{\varepsilon^2} +$$

$$+ P\{\int_0^\infty |a_n(s)|_H^2 \, d\langle (I-\Pi^k)M\rangle_s > \delta\} + \frac{16\delta}{\varepsilon^2} +$$

$$\le P\{\int_0^\infty |\Pi^k a_n(s)|_H^2 \, d\langle M\rangle_s > \delta\} + \frac{32\delta}{\varepsilon^2} +$$

$$+ P\{[\sup_t |a_n(t)|_H <(I-\Pi^k)M>_\infty] > \delta\}. \tag{12}$$

From (11)-(12) it follows the statment of the lemma 4.

Define

$$A_n(u) = A_{\tau_j^{(n)}}, \ u \in [\tau_j^{(n)}, \tau_{j+1}^{(n)}), \ j = 0,1,\dots,l_n-1.$$

Lemma 5. We have that for each $\varepsilon > 0$

$$\lim_{n\to\infty} P\{\sup_t |\int_0^t A_n(u)dZ_u - \int_0^t A_{u-}dZu| > \varepsilon\} = 0 .$$

This statement follows from the lemma 3 and the fact that for each $u > 0$ $A_n(u) - A_{u-} \to 0$ weakly in H as $n \to \infty$.

Denote $U = \{\tau_j^{(n)}, 0\le j\le l_n, n\ge 1$,

$$K_n(t) = \sum_{\tau_{j+1}^{(n)}\le t} |A_{\tau_{j+1}^{(n)}} - A_{\tau_j^{(n)}}|_H^2 ,$$

$$K_t = \sum_{s\le t} |\Delta A_s|_H .$$

Lemma 6. For $t \in U$ in probability

$$\lim_{n\to\infty} K_n(t) = K_t.$$

P r o o f . From the equality (6) it follows the existence of the limit in probability $\lim_{n\to\infty} K_n(t) = \overline{K}_t$. So as far

$$\sum_{\tau_{j+1}^{(n)}\le t} |A_{\tau_{j+1}^{(n)}} - A_{\tau_j^{(n)}}|_H^2 = \sum_{k=1}^{\infty} \sum_{\tau_{j+1}^{(n)}\le t} |(A_{\tau_{j+1}^{(n)}} - A_{\tau_j^{(n)}}, e_k)_H|^2 \tag{13}$$

and

$$\lim_{n\to\infty} \sum_{\tau_{j+1}^{(n)}\le t} |(A_{\tau_{j+1}^{(n)}} - A_{\tau_j^{(n)}}, e_k)_H|^2 = \sum_{s\le t} |(\Delta A_s, e_k)_H|^2, \tag{14}$$

then from (13)-(14) and the Fatou lemma we conclude that P - a.e. $K_t \le \overline{K}_t$, $t \ge 0$.

Now we have to prove the reverse inequality. We have

$$\sum_{\tau^{(n)}_{j+1}\le t} |A_{\tau^{(n)}_{j+1}}-A_{\tau^{(n)}_j}|^2_H = \sum_{\tau^{(n)}_{j+1}\le t} \{|\Delta A_{\tau^{(n)}_{j+1}}|^2_H - |A_{\tau^{(n)}_{j+1}}-A_{\tau^{(n)}_j}|^2_H + \tag{15}$$

$$+2[(A_{\tau^{(n)}_{j+1}}-A_{\tau^{(n)}_j}, X_{\tau^{(n)}_{j+1}}-X_{\tau^{(n)}_j})_H +$$

$$(A_{\tau^n_{j+1}}-A_{\tau^{(n)}_j}, Z_{\tau^{(n)}_{j+1}}-Z_{\tau^{(n)}_j})_H]\}.$$

But as $n \to \infty$ P - a.e.

$$\sum_{\tau^{(n)}_{j+1}\le t} |(A_{\tau^{(n)}_{j+1}}-A_{\tau^{(n)}_j}, X_{\tau^{(n)}_{j+1}}-X_{\tau^{(n)}_j})_H| \le \int_0^\infty |X^{(1)}_n(s) -$$

$$- X^{(2)}_n(s)|_V d|A|_s + |(X_0, A_{\tau^{(n)}_1 -})_H| \to 0. \tag{16}$$

Further according to the corollary 2 and the assumption that $Z \in \mathcal{S}(H)$, as $n \to \infty$ P - a.e.

$$\sum_{\tau^{(n)}_{j+1}\le t} [2(A_{\tau^{(n)}_{j+1}}-A_{\tau^{(n)}_j}, Z_{\tau^{(n)}_{j+1}}-Z_{\tau^{(n)}_j})_H - |A_{\tau^{(n)}_{j+1}}-A_{\tau^{(n)}_j}|^2_H]$$

$$\le \sum_{\tau^{(n)}_{j+1}\le t} [|(Z_{\tau^{(n)}_{j+1}}-Z^{(k)}_{\tau^{(n)}_{j+1}})-(Z_{\tau^{(n)}_j}-Z^{(k)}_{\tau^{(n)}_j})|^2_H +$$

$$+ 2(A_{\tau^{(n)}_{j+1}}-A_{\tau^{(n)}_j}, Z^{(k)}_{\tau^{(n)}_{j+1}}-Z^{(k)}_{\tau^{(n)}_j})_H|] \to 0. \tag{17}$$

Thus from (15) - (17) we obtain that P - a.e. $\overline{K}_t \le K_t$, $t \ge 0$. Lemma 6 is proved.

From formula (6) and the lemmas proved above we have that P - a.e. for each $t \in I$

$$|X_t|_H^2 = |X_0|_H^2 + 2\int_0^t X_s dA_s + 2\int_0^t A_{s-} dZ_s + |Z_t|_H^2 - \\ - \sum_{s\leq t} |\Delta A_s|_H . \quad (18)$$

P - a.e., if $t \in \overline{I} \setminus I$, and

$\sigma^{(n)} = \inf\{\tau_{j+1}^{(n)} : \tau_j^{(n)} \leq t < \tau_{j+1}^{(n)}\} \downarrow t$, then

$$|X_{\sigma}(n) - X_{\sigma}(n+p)|_H^2 = |X_{\sigma}(n+p)|_H^2 - |X\ (n)|_H^2 - \\ - 2(X_{\sigma}(n), X_{\sigma}(n+p) - X_{\sigma}(n))_H \to 0,$$

because according to (18) as $n \to \infty$ P - a.e.

$$|X_{\sigma}(n+p)|_H^2 - |X_{\sigma}(n)|_H^2 \to 0$$

and

$$|(X_{\sigma}(n), X_{\sigma}(n+p) - X_{\sigma}(n))_H| \leq |(X_{\sigma}(n), Z_{\sigma}(n+p) - Z_{\sigma}(n))_H| + \\ + |\int_{\sigma^{(n)}}^{\sigma^{(n+p)}} (X_n^{(2)}(s) - X_s) dA_s| + |\int_{\sigma^{(n)}}^{\sigma^{(n+p)}} X_s dA_s| \leq \text{const}\ Z_{\sigma}(n+p) \\ -Z_{\sigma}(n)|_H + \int_0^{\infty} |X_2^{(2)}(s) - X_s|\ d|A|_s + \int_{\sigma^{(n)}}^{\sigma^{(n+p)}} |X_s|_V d|A|_s \to 0.$$

Thus P - a.e. $X_{\sigma}(n) \to X_t$ in the strong topology of H .

Analogously P - a.e. for $t \in \overline{I} \setminus I$ and $\hat{\sigma}^{(n)} = \max\{\tau_j^{(n)} : \tau_j^{(n)} < t \leq \tau_{j+1}^{(n)}\} \uparrow t$ we have that $X_{\sigma}(n) \to X_{t-}$ in the strong topology of H. If $t,s \in \overline{I}^c$ and are contained in the same open interval, then $|X_t-X_s|_H^2 = |Z_t-Z_s|_H^2$. From what was said above it follows that X P - a.e. has the right continous paths with left limits. Together with the formula (18) it proves the formula (4).

Now we shall prove the formula (5). We have

$$X_t = X_o + F_t + G_t + \int_o^t \int_{H\setminus\{o\}} (x-h(x))p(du,dx),\ t \geq 0,$$

where

$$F_t = A_t + \sum_{s\leq t} \Delta Z_s \chi\{|\Delta Z_s|_H>1\} \int_o^t \int_{H\setminus\{o\}} (x-h(x))p(du,dx)$$

and

$$G_t = Z_t - \sum_{s\leq t} \Delta Z_s \chi\{|\Delta Z_s|_H>1\}.$$

It is clear that $\sup_t|\Delta F_t|_H \leq \text{const}$, $F \in \mathcal{A}_{loc}(V')$, $G \in \mathcal{S}p(H)$. The following unique decompositions hold (see, e.g. [24]):

$$F_t = \tilde{F}_t + L_t,\ G_t = \tilde{G}_t + R_t,\ t \geq 0,$$

where $F \in \mathcal{A}_{loc}(V') \cap \mathcal{P}$, $L \in \mathcal{M}_{loc}(V')$,

$\tilde{G} \in \mathcal{A}_{loc}(H) \cap \mathcal{P}$, $R \in \mathcal{M}^2_{loc}(H)$.

According to the formula (4) P - a.e. $\sum_{s\leq t} |\Delta X_s|_H^2<\infty, t\geq 0$.

Therefore the integral $\int_o^t \int_{H\setminus\{o\}} h(x)q(du,dx),\ t \geq 0,$ is defined and $X^c \in \mathcal{M}^c_{loc}(V')$, where

$$X_t^c = X_t-X_o-\int_o^t \int_{H\setminus\{o\}} (x-h(x))p(du,dx)-\int_o^t \int_{H\setminus\{o\}} h(x)q(ds,dx)- \\ - \tilde{F}_t - \tilde{G}_t,\ t \geq 0.$$

We have that

$$X_t^c + \int_0^t \int_{H\setminus\{o\}} h(x)q(du,dx) = L_t + R_t^c + R_t^d,\ t \geq 0,$$

where R^c and R^d are continuous and purely discontinuous parts of R, respectively. For each $v \in V$ we shall have

$$<X_t^c,v> + \int_o^t \int_{H\setminus\{o\}} (h(x),v)_H\ q(du,dx) = <L_t,v> +$$

$$+ (R_t^c,v)_H + (R_t^d,v)_H,\ t \geq 0 . \qquad (20)$$

As far as $<L_t,v>,\ t \geq 0,$ is a purely discontinuous local martingale, then from (19) - (20) we find that

$$L_t + R_t^d = \int_0^t \int_{H\setminus\{o\}} h(x)q(du,dx),\ t \geq 0,$$

and $R_t^c = X_t^c,\ t \geq 0,$ i.e. $L \in M^2_{loc}(H)$. The formula (5) now follows from (19), if we denote $\alpha_t^h = \tilde{F}_t + \tilde{G}_t,\ t \geq 0$. Theorem 2 is proved.

Remark 1. Under the assumptions of theorem 2 we can correctly define the integral $\int_0^t X_{s-}d\alpha_s^h,\ t \geq 0,$ by means of the following formula:

$$\int_0^t X_{s-}d\alpha_s^h = \int_0^t X_s dC_s - \sum_{s\leq t} (\Delta X_s,\Delta C_s)_H + \int_0^t X_{s-}dS_s,\ t \geq 0.$$

Remark 2. The triplet (α^h,β,Π) where β is defined by the equality

$$<(X^c,v)_H>_t = (v,\beta_t v)_H,\ t \geq 0,\ v \in V ,$$

is called the triplet of the $(P,\mathbb{F})$-predictable

characteristics of the process X . The class of processes X , permitting the decomposition (5), we shall denote by $S(V,H,P,\mathbb{F})$. Note that for functions h_1 and h_2, satisfying analogous assumptions to h, we have the equality:

$$\alpha_t^{h_2} = \alpha_t^{h_1} + \int_0^t \int_{H\setminus\{o\}} (h_1(x)-h_2(x))\Pi(ds,dx),\ t \geq 0 .$$

3. Weak convergence of semimartingales with values in a rigged Hilbert space

Now we shall consider the space $X = D_{[0,\infty)}(H) \cap L^2_{loc}(V,N)$ with the family of σ-algebras $\mathcal{B}_t = \sigma(X_s, s\leq t)$, $t \geq 0$, where $X_t(\hat{\omega}) = \hat{\omega}(t)$, $\hat{\omega} \in X$, and the probability measure $\hat{P}$ on $\mathcal{B}(X)$, assuming that it is uniquely defined by the given restriction $\hat{P}|\mathcal{B}_o = \mu_o$ and the assumption that $X \in S(V,H,\hat{P},\mathbb{B}_+)$ with the given triplet $(\hat{\alpha}^h,\hat{B},\hat{\Pi})$ of the $(\hat{P},\mathbb{B}_+)$-predictable characteristics and $|d\hat{\alpha}_t^h|_{V'} \leq (1+|X_t|_V)dN_t$.

Denote

$$\hat{\Gamma}_v^h(t,\omega) = (v,B_t v)_H + \int_{H\setminus\{o\}} (h(x),v)_H^2\ \hat{\Pi}(\hat{\omega},[o,t]\times dx) -$$

$$- \sum_{s\leq t} \Big(\int_{H\setminus\{o\}} (h(x),v)_H\ \hat{\Pi}(\hat{\omega},\{s\}\times dx)\Big)^2,\ t\geq 0,\ v\in V,$$

$$M_t^\varphi = p_t(\varphi) - \hat{\Pi}_t(\varphi),\ \varphi \in \hat{C}^+(H\setminus\{o\}) ,$$

$$Y_t^h = X_t - X_o - \hat{\alpha}_t^h - \int_0^t \int_{H\setminus\{o\}} (x-h(x))p(ds,dx),\ t\geq 0;$$

$\hat{C}^+(H\setminus\{o\})$ is a space of continuous bounded functions which are equal to zero in the neighbourhood of zero,

$$\nu_t(\varphi) = \int_{H\setminus\{0\}} \varphi(x)\nu([0,t]xdx).$$

It is easy to check that $X \in S(V,H,\hat{P},\mathbb{B}_+)$ with the triplet $(\hat{\alpha}^h,\hat{B},\hat{\Pi})$ of the $(\hat{P},\mathbb{B}_+)$-predictable characteristics iff the families $\{(Y^h_t,v)_H,\ t\geq 0,\ v\in V\}$, $\{(Y^h_t,v)^2_H - \hat{\Gamma}^h_v(t),\ t\geq 0,\ v\in V\}$ and $M^\varphi_t,\ t\geq 0,\ \varphi\in\hat{C}^+(H\setminus\{0\})\}$ are the families of the $(\hat{P},\mathbb{B}_+)$-local martingales.

We shall assume that $X_n \in S(V,H,P_n,F_n)$ with the paths in the space X, the triplets $(\alpha^{h,(n)},B^{(n)},\Pi^{(n)})$ of the $(P_n,\mathbb{F}_n)$-predictable characteristics and initial distributions $\mu_0^{(n)}$, $n\geq 1$.

An immediate consequence of the theorem 1 is the following theorem.

Theorem 3. Assume that:

1) the sequence $\{\hat{P}_n,\ n\geq 1\}$ is weakly relative compact;
2) $\mu_0^{(n)} \Rightarrow \mu_0$ as $n\to\infty$;
3) for each $t>0$, $v\in V$ and $\varphi\in\hat{C}^+(H\setminus\{o\})$ $(\hat{\alpha}^h_t(\cdot),v)_H$, $(v,\hat{B}_t(\cdot)v)_H$ and $\hat{\Pi}_t(\cdot,\varphi)$ are continuous in the topology of X;
4) for each $v\in V$, $\varphi\in\hat{C}^+(H\setminus\{0\})$ and $T>0$ there exist a sequence $\{S_n,\ n\geq 1\}\in\Sigma(T)$ and $G\in K$ such that

$$\sup_{n\geq 1} E_n[G(\Gamma_v^{h,(n)}(S_n) + \Pi_{S_n}^{(n)}(\varphi))] < \infty;$$

5) for each $\varepsilon>0$, $t>0$, $v\in V$ and $\varphi\in\hat{C}^+(H\setminus\{0\})$

$$\lim_{n\to\infty} P_n\{|(\alpha_t^{h,(n)},v)_H - (\hat{\alpha}^h_t(X_n),v)_H| > \varepsilon\} = 0,$$

$$\lim_{n\to\infty} P_n\{|\Gamma_V^{h,(n)}(t) - \hat{\Gamma}_V^h(t,X_n)| > \varepsilon\} = 0 ,$$

$$\lim_{n\to\infty} P_n\{|\Pi_t^{(n)}(\varphi) - \hat{\Pi}_t(X_n,\varphi)| > \varepsilon\} = 0 .$$

Then $\hat{P}_n \Rightarrow \hat{P}$ as $n \to \infty$.

At the end of the paper we shall give a sketch of the proof of a criterion for the weakly relative compactness of $\{\hat{P}_n, n \geq 1\}$.

Denote $\tilde{\Sigma}(T) = (T_{n,j})_{n,j\geq 1} : T_{n,j} \in \mathcal{T}(\mathbb{F}_n)$,

$T_{n,j} \leq T, \lim_{j\to\infty} \overline{\lim_{n\to\infty}} P_n\{T_{n,j} < T\} = 0\}$,

$\bar{\Pi}^k = I - \Pi^k,\ k \geq 1,\ \bar{\Pi}^0 = I.$

Theorem 4. Let the following assumptions hold:

1) $\{\mu_0^{(n)}, n\geq 1\}$ is weakly relative compact;

2) for each $T > 0$ and $K > 0$ there exist $(T_{n,j}) \in \tilde{\Sigma}(T)$, the basis $\{e_k, k\geq 1\}$ of H, consisting of the elements from V, $\lambda_{n,j} \in A^+ \cap P$ and $A_{n,k} \in A^+ \cap P$ such that for each $\varepsilon > 0$

$$\lim_{l\to\infty} \overline{\lim_{n\to\infty}} P_n\{\Pi^{(n)}([0,T]\times\{x:|x|_H>l\} > \varepsilon\} = 0 ,$$

$$\sup_{n,j,k} |X_{n,j}(\infty) + A_{n,k,j}(\infty)| \leq \text{const},$$

$$\lim_{k\to\infty} \overline{\lim_{n\to\infty}} P_n\{A_{n,k,j}(\infty) > \varepsilon\} = 0$$

and on the set $]0,T_{n,j}]$

$$2\langle\bar{\Pi}^k X_n(s-), d\alpha_s^{(n)}\rangle + 2(\bar{\Pi}^k X_n(s-), \int_{1<|x|_H\leq K} x\, \Pi^{(n)}(ds,dx))_H^+$$

$$+ d\langle\bar{\Pi}^k X_n^c\rangle_s + \int_{|x|_H\leq K} |\Pi_x^k|_H^2\, \Pi^{(n)}(ds,dx) + \varepsilon_k |\bar{\Pi}^k X_n(s-)|_V^2 dN_s$$

$\leq |X_n(s-)|_H^2 dX_{n,j}(s) + dA_{n,k,j}(s),$

where $\varepsilon_o > 0$, $\varepsilon_k \geq 0$, $k \geq 1$;

3) for each $T_n \in \mathcal{T}(\mathbb{F}_n)$, $T_n \leq T$, $\delta_n \downarrow 0$, $j \geq 1$ and $\varepsilon > 0$

$$\lim_{n\to\infty} P_n\{|(\alpha^{(n)}_{T_n+\delta_n}, e_j)_H - (\alpha^{(n)}_{T_n}, e_j)_H| + \Gamma^{h,(n)}_{e_j}(T_n+\delta_n) - \Gamma^{h,(n)}_{e_j}(T_n) + \Pi^{(n)}((T_n, T_n+\delta_n]\times\{x: |x|_H > 1\}) > \varepsilon\} = 0.$$

Then $\{\hat{P}_n, n\geq 1\}$ is weakly relative compact.

P r o o f . From assumption 2) we have that

$$\lim_{l\to\infty} \overline{\lim_{n\to\infty}} P_n\{\max_{u\leq T} |\Delta X_n(u)|_H > l\} = 0 .$$

Therefore, denoting $S_{n,l} = \inf\{t: |\Delta X_n(t)|_H > l\}$ we shall have that $(S_{n,l}) \in \tilde{\Sigma}(T)$. From the definition of $\tilde{\Sigma}(T)$ and the general form of compact subsets in the spaces $D_{[0,\infty)}(H)$ and $L^2_{loc}(V,N)$ it is clear that it is enough to prove for each $T > 0$, $j \geq 1$ and $l \geq 1$ the weakly relative compactness of the sequence of measures, corresponding to the processes $X_{n,T_{n,j} \wedge S_{n,l}-}$ (we shall denote them later by the same letter X_n), where $f_{S-}(t) = f(t)$ for $t < S$ and $= f(S-)$ for $t \geq S$. Besides, according to 1), we can assume that $\sup|X_n(0)|_H \leq$ const and $\lim_{k\to\infty} \sup_n E_n|\overline{\Pi}^k X_n(0)|_H = 0$.

Using the formula (4) we have that

$$|\overline{\Pi}^k X_n(t)|_H^2 = |\overline{\Pi}^k X_n(0)|_H^2 + 2\int_0^t \overline{\Pi}^k X_n(s-)\, d\alpha_n(s) +$$

$$+ 2\int_0^t \int_{|x|_H>1} (\overline{\Pi}^k X_n(s-),x)_H p_n(ds,dx) + 2 \int_0^t \overline{\Pi}^k X_n(s-)dX_n^s(s)$$

$$+ 2\int_0^t \int_{|x|_H\le 1} (\overline{\Pi}^k X_n(s-),x)_H \, q_n(ds,dx) + \langle\overline{\Pi}^k X_n^c\rangle_t +$$

$$+ \int_0^t \int_{H\setminus\{0\}} |\overline{\Pi}^k x|_H^2 \, P_n(ds,dx) \, .$$

From here by means of the lemma 2 of the paper [29] and the assumption 2) we obtain that

$$\lim_{k\to\infty} \overline{\lim_{n\to\infty}} \sup_{\sigma\in T(\mathbb{F}_n)} E_n|\overline{\Pi}^k X_n(\sigma)|_H^2 = 0$$

and for each $T > 0$

$$\sup_n E_n[\int_0^T |X_n(s-)|_V^2 dN_s] < \infty \, .$$

At last from the assumption 3) we have that for each $\varepsilon > 0$ and $k \ge 1$

$$\lim_{n\to\infty} P_n\{|(X_n(T_n+\delta_n),e_k)_H - (X_n(T_n),e_k)_H| > \varepsilon\} = 0.$$

To complete the proof of the theorem 4 now it is enough to apply the criterion from the paper [30] and the known criterion of weakly relative compactness of probability measures in H (see [31]).

B. Grigelionis, R. Mikulevičius
Institute of Mathematics and Cybernetics
Lithuanian Academy of Sciences
University of Vilnius

Vilnius, Lithuania
U.S.S.R.

References

1. Galavotti, G.; Jona-Lasinio, G.: Limit theorems for multidimensional markovian processes. - Comm. Math. Phys., 1975, vol. 41, p. 301-307

2. Viot, M.: Solutions faibles d'équations aux dérivées partielles stochastiques non lineaires. - These doct. sci. Univ. P. et M. Curie, Paris, 1976.

3. Martin-Löf, A.: Limit theorems for motion of a Poisson system of independent markovian particles with high density. - Z. Wahr.verw. Geb., 1976, B. 34, S. 205-223

4. Dawson D. A. Stochastic evolution equations and related measure processes. - J. Multivariate Anal., 1975, vol. 5, No. 1, p. 1-52

5. Dawson, D. A.: Critical measure diffusion processes. - Z. Wahr.verw. Geb., 1977, B. 40, S. 125-145

6. Dawson, D. A.: Stochastic measure processes. - Stochastic Nonlinear Systems, ed. L. Arnold, R. Lefever, Springer, 1981, p. 185-199

7. Dawson, D. A.; Hochberg, K. J.: Wandering random measures in the Fleming - Viot model. - Carleton Math. Lecture Notes No. 33, 1981

8. Dawson, D. A.: Critical dynamics and fluctuations for a mean field model of cooperative behaviour. - Carleton Math. Lecture Note No. 33, 1981

9. Holley, R.; Stroock, D. W.: Generalized Ornstein - Uhlenbeck processes and infinite particle branching brownian motions. - RIMS Kyoto Univ., 1978, vol. 14, p. 741-788

10. Holley, R.; Stroock, D. W.: Central limit phenomena of various interacting systems. - Ann. of Math., 1979, vol. 110, p. 333-393

11. Holley, R.; Stroock, D.: Generalized Ornstein - Uhlenbeck processes as limits of interacting systems. - Proc. LMS Durham Symp., p. 152-168; Lecture Notes in Math. 851, Springer, 1981

12. Vishik, M. I.: Komech, A. I.; Fursikov A. V.: Mathematical problems of hydromechanics. - Uspechi Math. Nauk., 1979, vol. 34, No. 5, p. 135-210

13. Arnold, L.; Theodosopulu M.: Deterministic limit of the stochastic model of chemical reaction with diffusion. - Adv. Appl. Probab., 1980, vol. 12, p. 367-379

14. Arnold, L.: Mathematical models of chemical reactions. - Stochastic systems, Dordrecht, 1981

15. Kurtz, T. G.: Approximation of population processes. - CBMS - NSF Regional Conference series in Applied Mathematics, SIAM, 1981.

16. Tanaka, H.; Hitsuda, M.: Central limit theorem for a simple diffusion model of interacting particles. - Hirosima Math. J., 1981, vol. 11, p. 415-423

17. Hitsuda M. Central limit theorem for a simple interacting diffusion model and S'-valued processes. - Lecture Notes in Math., 1021, Proc. IV USSR-Japan Symp. Probab. Th. and Math. Stat., p. 238-242, Springer, 1983.

18. Kotelenez, P.: Law of large numbers and central limit theorem for chemical reaction with diffusion. - Dr. dissertation Univ. Bremen, 1982

19. Bouc R.; Pardoux, E.: Asymptotic analysis of PDEs with wideband driving noise. - Publ. Math. Appl. Marseille-Toulon, 83-4, 1983

20. Fouque, J.-P.: La convergence en loi pour les processus a valeurs dans un espace nucleaire. - Preprint, 1983, Paris

21. Bourbaki, N.: Integration, ch. IX. - Hermann, Paris

22. Grigelionis, B.; Mikulevičius, R.: On stably weak convergence of semimartingales and point processes. - Theory of Probab. and Appl., 1983, vol. XXVIII, No. 2, p. 320-332

23. Jacod, J.: Calcul stochastique et problemes de martingales. - Lecture Notes in Math. 714, Springer Verlag, 1979

24. Metivier, M.; Pellaumail, J.: Stochastic integration. - Academic Press, New York - London, 1980

25 Metivier, M.: Semimartingales. - W. de Gruyter, Berlin - New York, 1982

26. Krylov, N.V.; Rozovskii, B.L.: On stochastic evolution equations. - Modern Problems in Math., vol. 14, VINITI, M., 1979, p. 71-146

27. Gyöngy, I.; Krylov, N.V.: On stochastic equations with respect to semimartingales II, Ito formula in Banach spaces. - Stochastics, 1982, vol. 6, No. 3 + 4, p. 153-173

28. Lenglart E. Relation de domination entre deux processus. - Ann. Inst. H. Poincaré, 1977, vol. 13, p. 171-179

29. Grigelionis, B.; Mikulevičius, R.: Stochastic evolution equations and densities of the conditional distributions. - Lecture notes in Control and Inform. Theory, 49, Springer, 1983

30. Aldous, D.: Stopping times and tightness. - Ann. Probab., 1978, vol. 6, No. 2, p. 335-340

31. Parthasarathy, K. R.: Probability measures on metric space. - Academic Press, N.Y. - L., 1967

Michel Metivier

TIGHTNESS OF SEQUENCES OF HILBERT VALUED MARTINGALES

Abstract

This paper gives a sufficient condition for the tightness of a sequence $(M^n)_{n \in \mathbb{N}}$ of Hilbert valued martingales. This result applies directly to some situations of "accompanying martingales" as considered for example by L. Arnold, M. Theodosopulu and P. Kotelenez.

1. INTRODUCTION

Let us recall that if M is a $\mathbb{H}$-valued martingale (on a stochastic basis $(\Omega, (\mathscr{F}_t)_{t \in [0,T]}, P)$), $\mathbb{H}$ being a separable Hilbert-space, there is a unique process denoted by $\langle\!\langle M \rangle\!\rangle$ (see [6] chapter 4) with the following properties : $\langle\!\langle M \rangle\!\rangle$ is the $\mathbb{H} \hat{\otimes}_1 \mathbb{H}$-valued process (matrix-valued process) such that $M \otimes M - \langle\!\langle M \rangle\!\rangle$ is a martingale, with $\langle\!\langle M \rangle\!\rangle$ predictable and with finite variation. Let us recall that if (h_i) is an orthonormal basis of $\mathbb{H}$, the elements of $\mathbb{H} \hat{\otimes}_1 \mathbb{H}$ are of the form $\hat{y} := \sum_{i,j \in \mathbb{N}} \lambda_{ij} h_i \otimes h_j$ with $\|\hat{y}\|_1 = \sum |\lambda_{ij}| < \infty$, and $\mathbb{H} \hat{\otimes}_1 \mathbb{H}$ is a Banach space for the norm $\|\hat{y}\|_1$. This space is included in the Hilbert-Schmidt tensor product

$$\mathbb{H} \hat{\otimes}_2 \mathbb{H} := \{\hat{y} : \hat{y} = \sum_{i,j} \lambda_{i,j} h_i \otimes h_j , \sum_{ij} |\lambda_{ij}|^2 < \infty\} .$$

L. Arnold and P. Kotelenez (eds.), Stochastic Space-Time Models and Limit Theorems, 217–226.

The space $\mathbb{H} \hat{\otimes}_2 \mathbb{H}$ is a Hilbert space with $h_i \otimes h_j$ as an orthonormal basis and norm $\|\hat{y}\|_2 = (\sum_{ij} |\lambda_{ij}|^2)^{\frac{1}{2}}$. The injection from $\mathbb{H} \hat{\otimes}_1 \mathbb{H}$ into $\mathbb{H} \hat{\otimes}_2 \mathbb{H}$ is continuous.

We denote by $\lessdot M \gtrdot$ the real valued increasing process :

$$(1.1) \qquad \lessdot M \gtrdot_t := \text{trace} \ll M \gg_t := \sum_i \ll M \gg_t^{ii} .$$

In most practical problems the process $\ll M \gg$ has an integral representation of the type

$$(1.2) \qquad \ll M \gg_t = \int_0^t \Phi(s,\omega)\,ds$$

where Φ is a $\mathbb{H} \hat{\otimes}_1 \mathbb{H}$-valued process.

The theorem we present here (theorem 2.2) gives a tightness condition for a sequence $(M^n)_{n \in \mathbb{N}}$ of martingales involving only the processes $\ll M^n \gg$ and which applies very easily when these processes have the form (1.2).

Each process (M^n) is assumed to be defined on a stochastic basis $(\Omega^n, (\mathcal{F}_t^n)_{t \in [0,T]}, P^n)$.

2. A TIGHTNESS CONDITION

An immediate extension of a result due to Rebolledo (see [9] or [8] for a short exposition) is the following.

2.1 Theorem (Rebolledo)

Let $(M^n)_{n \in \mathbb{N}}$ be a sequence of $\mathbb{H}$-valued right continuous martingales. We assume the following two properties [A] and [T] for this sequence :

[A] For every ε , $\eta > 0$ there exists $\delta > 0$ such that for every sequence (τ^n) of stopping times (τ^n is a stopping time on Ω^n with respect to the filtration $(\mathscr{F}^n_t)_{t \in [0,T]}$)

$$\limsup_{n \to \infty} P^n \{ \langle\!\langle M^n \rangle\!\rangle_{\tau^n+\delta} - \langle\!\langle M^n \rangle\!\rangle_{\tau} > \eta \} \leqslant \varepsilon$$

[T] For every $t \in [0,T]$ and every $\varepsilon > 0$ there exists a compact set C in $\mathbb{H}$ such that

$$\sup_n P^n \{ M^n_t \notin C \} \leqslant \varepsilon \ .$$

Then, denoting by $\widetilde{P}^n$ the law of X^n in the Skorokhod space $\widetilde{\Omega} := \mathbb{D}(0,T\,;\mathbb{H})$ of the process M^n , the sequence $(\widetilde{P}^n)_{n \in \mathbb{N}}$ is tight.

What we give now is a sufficient condition for [T] to be true. As said previously, this condition involves only the processes $\langle\!\langle M^n \rangle\!\rangle$.

2.2 Theorem

If the sequence $(\langle\!\langle M^n \rangle\!\rangle^{\frac{1}{2}})_{n \in \mathbb{N}}$ of $\mathbb{H} \hat{\otimes}_2 \mathbb{H}$-valued processes satisfies [T] the same condition holds for the process $(M^n)_{n \in \mathbb{N}}$.

Proof. By assumption, for each t in $[0,T]$ there exists a compact subset $\hat{C}$ of $\mathbb{H} \hat{\otimes}_2 \mathbb{H}$ such that

$$\sup_n P^n \{ \langle\!\langle M^n \rangle\!\rangle_t \notin \hat{C} \} \leqslant \varepsilon \ . \tag{2.2.1}$$

Being compact the subset $\hat{C}$ is bounded and such that for every $\rho > 0$ there exists a finite dimensional subspace $\hat{\mathbb{F}}$ of $\mathbb{H} \hat{\otimes}_2 \mathbb{H}$ such that $\sup_{y \in \hat{C}} \| y - \Pi_{\hat{\mathbb{F}}}(y) \|^2 \leq \rho$, $\Pi_{\hat{\mathbb{F}}}$ denoting the orthogonal projection on $\hat{\mathbb{F}}$.

Our first claim is that the finite dimensional subspace $\hat{\mathbb{F}}$ can be taken of the form $\mathbb{H}_\rho \otimes \mathbb{H}_\rho$ where $\mathbb{H}_\rho$ is a finite dimensional subspace of $\mathbb{H}$. In fact if $\sum_{i,j} \xi^n_{ij} h_i \otimes h_j$ $n=1,\ldots,K$ is an orthonormal basis of a subspace $\hat{\mathbb{F}}$ of $\mathbb{H} \hat{\otimes}_2 \mathbb{H}$ such that $\sup_{y \in \hat{C}} \| y - \Pi_{\hat{\mathbb{F}}}(y) \|^2 \leq \rho/2$ and if we consider the subspaces $\hat{\mathbb{F}}_J$ generated by the vectors $\{ \sum_{i,j \in J} \xi^n_{ij} h_i \otimes h_j \,,\ n=1,\ldots,K\}$ for each finite subset J of $\mathbb{N}$

$$\lim_{J \uparrow} \| y - \Pi_{\hat{\mathbb{F}}_J}(y) \| = \| y - \Pi_{\hat{\mathbb{F}}}(y) \| \ .$$

The compactness of $\hat{C}$ and the continuity of $y \to \Pi_{\hat{\mathbb{F}}}(y)$ implies the existence of J such that

$$\text{(2.2.2)} \qquad \sup_{y \in \hat{C}} \| y - \Pi_{\hat{\mathbb{F}}_J}(y) \| \leq \rho \ .$$

But $\hat{\mathbb{F}}_J = \mathbb{H}_J \otimes \mathbb{H}_J$ where $\mathbb{H}_J$ is the subspace of $\mathbb{H}$ generated by $\{h_i : i \in J\}$.

Our second step will be to prove that for every $\varepsilon > 0$ and $\beta > 0$ we can find $\alpha > 0$ and a finite dimensional subspace $\mathbb{H}_\rho$ such that

$$\text{(2.2.3)} \qquad P^n\{\| M^n_t \| > \alpha\} \leq \varepsilon$$

and

(2.2.4) $\quad P^n\{\|M_t^n - \Pi_{H_\rho} M_t^n\|^2 > \beta\} \leqslant \varepsilon$

It is easily seen that these properties will imply the hypothesis [T] for the sequence $(M_t^n)_{n\in\mathbb{N}}$ of $\mathbb{H}$-valued random variables.

To derive (2.2.3) we use the following consequence of a lemma of Lenglart (see [5] or [8]) : for every $b > 0$, $a > 0$ and every stopping time τ^n :

(2.2.5) $\quad P^n\{\sup_{s\leqslant\tau^n} \|M_s^n\|^2 > b^2\} \leqslant \frac{1}{b^2} E\{\langle M^n\rangle_{\tau^n} \wedge a\} + P\{\langle M^n\rangle_{\tau^n} \geqslant a\}$

But, since $\langle M^n\rangle_t = \text{trace} \langle\langle M^n \rangle\rangle_t = \|\langle\langle M^n\rangle\rangle_t^{\frac{1}{2}}\|_{H.S}$, if we take $b := \sup_{y\in\hat{C}} \|y\|^2_{H.S}$ we see that

$$P^n\{\|M_t^n\| > \alpha\} \leqslant b/a^2 + \varepsilon/2 .$$

Therefore α can be chosen for (2.2.3) to hold. To prove (2.2.4) let us take $\rho \leqslant \varepsilon/2\ \beta^2$ and consider the finite dimensional subspace F_ρ as considered above. Let G_ρ be the orthogonal subspace of $\mathbb{H}_\rho$, call Π_1 the orthonormal projection on G_ρ and Π_2 the orthogonal projection on $\mathbb{H}_\rho$. To simplify we set $Q := \langle\langle M\rangle\rangle_t$. Let us assume we are able to prove

(2.2.6) $\quad \langle \Pi_1 M\rangle_t \leqslant \|Q^{\frac{1}{2}} - \Pi_{H_\rho\otimes H_\rho} Q^{\frac{1}{2}}\|^2_{H.S}$

Applying again and using the fact that

$$P\{\| Q^{\frac{1}{2}} - \Pi_{H_\rho \otimes H_\rho} Q^{\frac{1}{2}} \|^2_{H.S} > \rho\} < \varepsilon/2$$

we obtain

$$P\{|\Pi_1 M|_t > \beta\} \leqslant \rho/\beta^2 + \varepsilon/2 .$$

From the choice of ρ we have

$$P\{|\Pi_1 M|_t > \beta\} \leqslant \varepsilon .$$

This is (2.2.3). The theorem will follow from the proof of (2.2.5). This is done in the next lemma.

2.3 Lemma

Let $Q := \langle\!\langle M \rangle\!\rangle_t$. Let $\mathbb{H}_1 \oplus \mathbb{H}_2$ be an orthogonal decomposition of $\mathbb{H}$ and $\Pi_{\mathbb{H}_2 \hat{\otimes} \mathbb{H}_2} Q^{\frac{1}{2}}$ the orthogonal projection of $Q^{\frac{1}{2}}$ on $\mathbb{H}_2 \hat{\otimes}_2 \mathbb{H}_2$. Let Π_1 be the orthogonal projection on $\mathbb{H}_2$. Then

$$\langle\!\langle \Pi_1 M \rangle\!\rangle_t \leqslant \| Q^{\frac{1}{2}} - \Pi_{\mathbb{H}_2 \hat{\otimes}_2 \mathbb{H}_2} Q^{\frac{1}{2}} \|^2_{H.S}$$

Proof. Call $\mathcal{Q}$ the operator in $\mathcal{L}(\mathbb{H}\,;\mathbb{H})$ associated to Q .

(2.3.1) $\quad (\mathrm{Proj}_{\mathbb{H}_i \hat{\otimes}_2 \mathbb{H}_j} Q^{\frac{1}{2}}) = \Pi_i \circ Q^{\frac{1}{2}} \circ \Pi_j \qquad i,j = 1,2.$

Let us write $\widetilde{Q}$ for the Hilbert-Schmidt operator associated to $Q \in \mathbb{H} \hat{\otimes}_2 \mathbb{H}$.

Note first that

(2.3.2) $\quad \langle\!\langle \Pi_2 M \rangle\!\rangle_t = \mathrm{trace}\, \Pi_2 \circ \widetilde{Q} \circ \Pi_2 .$

Then

$$\| \Pi_2 \circ \widetilde{Q}^{\frac{1}{2}} \circ \Pi_2 \|^2_{H.S} \leqslant \| \Pi_2 \circ \widetilde{Q}^{\frac{1}{2}} \|^2_{H.S} = \mathrm{trace}(\Pi_2 \circ \widetilde{Q}^{\frac{1}{2}} \circ \widetilde{Q}^{\frac{1}{2}} \circ \Pi_2) = \mathrm{trace}(\Pi_2 \circ \widetilde{Q} \circ \Pi_2) .$$

Therefore, from (2.3.2)

(2.3.3) $\| \Pi_2 \circ \tilde{Q}^{\frac{1}{2}} \circ \Pi_2 \|^2_{H.S} \leqslant \langle \Pi_2 M \rangle_t$.

Moreover, as a consequence of $\| M_t \|^2 = \| \Pi_1 M_t \|^2 + \| \Pi_2 M_t \|^2$

(2.3.4) $\langle\!\langle M \rangle\!\rangle_t = \langle \Pi_1 M \rangle_t + \langle \Pi_2 M \rangle_t$.

From

$$\langle\!\langle M \rangle\!\rangle_t = \sum_{i,j=1}^{2} \| \Pi_i \circ \tilde{Q}^{\frac{1}{2}} \circ \Pi_j \|^2_{H.S}$$

we obtain easily the inequality of the lemma. ∎

The following proposition gives a sufficient condition of wide use to check the condition of Theorem 2.2.

2.4 Proposition

Let us assume

(i) $\lim_{\rho \to \infty} \sup_n P^n\{ \langle\!\langle M^n \rangle\!\rangle_t > \rho \} = 0$

(ii) there exists a complete orthonormal system $(h_k)_{k \in \mathbb{N}}$ in $\mathbb{H}$ such that

$$\lim_{n \to \infty} \sup_n P^n\{ \sum_{k=m}^{\infty} (h_k ; \langle\!\langle \tilde{M}^n \rangle\!\rangle_t h_k) > \eta \} = 0 .$$

Then the sequence $(\langle\!\langle M^n \rangle\!\rangle_t^{\frac{1}{2}})_{n \in \mathbb{N}}$ is tight in $\mathbb{H} \hat{\otimes}_2 \mathbb{H}$.

Proof. It is easy to show the existence, for each $\varepsilon > 0$ and $\eta > 0$, of a finite dimensional subspace L_ε of $\mathbb{H} \hat{\otimes}_2 \mathbb{H}$ and a bounded set B_2 in $\mathbb{H} \hat{\otimes}_2 \mathbb{H}$ such that :

$$\sup_n P^n\{ \langle\!\langle M^n \rangle\!\rangle_t^{\frac{1}{2}} \notin B_\varepsilon \} \leqslant \varepsilon$$

and

$$\sup_n P^n\{\| \langle\!\langle M^n \rangle\!\rangle_t^{\frac{1}{2}} - \Pi(\langle\!\langle M^n \rangle\!\rangle_t^{\frac{1}{2}}\|_2 > \eta\} \leq \varepsilon$$

where Π denotes the projection operator on L_ε .

3. EXAMPLE OF APPLICATION

In [4] the following sequence of "accompanying martingales" $(M^n)_{n \in \mathbb{N}}$ is considered.

$\mathbb{H}$ is a Sobolev space $\mathbb{H}_{-q}$ (see [4] for definitions) which is the dual of $\mathbb{H}_q \hookrightarrow L^2[0,1]$. The martingales M^n are such that for every $u,v \in \mathbb{H}_q$:

$$(3.1) \qquad \langle u \,; \langle\!\langle M^n \rangle\!\rangle_t \, v\rangle = \int_0^t \langle u \,;\, a^n(X_s^n)\, v\rangle \, ds$$

where the processes X^n take their values in $L^2[0,1]$ and $g \rightsquigarrow a^n(g)$ is an operator on $L^2[0,1]$ such that for every $u,v \in \mathbb{H}_1$ and $g \in L_\infty[0,1]$:

$$(a^n(g)u;v) \leq K\, \varepsilon_n \, \|g\|_\infty \, \|u\|_1 \, \|v\|_1$$

If it is assumed, for example as in [3] , that on [0,T] the paths of X^n are bounded in $L_\infty[0,1]$, we obtain, using the fact that the imbedding $i : H_2 \rightsquigarrow H_1$ is Hilbert-Schmidt:

$$(3.2) \qquad \langle\!\langle \frac{1}{\sqrt{\varepsilon_n}} M^n \rangle\!\rangle_t = \int_0^t \frac{1}{\varepsilon_n} a^n(X_s^n)\, ds$$

with

$$(3.3) \qquad \text{trace } (\frac{1}{\varepsilon_n} a^n(X_s^n)) \leq K$$

and for any complete orthonormal basis in H_2

$$(3.4) \qquad \lim_{m\to\infty} \sum_{k=m}^{\infty} \langle h_k , \frac{1}{\varepsilon_n} \ll M^n \gg_t h_k \rangle \leqslant \lim_{m\to\infty} K\ T \sum_{k=m}^{n} \| h_k \|_1^2 = 0 \ .$$

Inequalities (3.3) and (3.4) easily imply the conditions of theorem 2.2 and proposition 2.3.

For more details on this type of applications the reader is referred to the forthcoming paper [7].

Acknowledgements. The author wants to thank the Center for Stochastic Processes of the University of North Carolina at Chapel Hill for supporting him under grant NO F49 620-82-C-0009 of the Air Force Office of Scientific Research during the summer 1983, while the results presented here were produced.

Professor Michel METIVIER
Centre de Mathématiques Appliquées -
Ecole Polytechnique
PALAISEAU CEDEX 91128
FRANCE

REFERENCES

[1] Aldous, D.: 1978, "Stopping times and tightness", Ann. of Prob. 6, 2, 335-340.

[2] Arnold, L.: 1981, "Mathematical models of chemical reactions" in M. Hazewinkel and J. Willems (ed.), Stochastic systems, Dordrecht.

[3] Arnold, L. and Theodosopulu, M.: 1980, "Deterministic limit of the stochastic model of chemical reactions with diffusion", Adv. Appl. Prob., 1 , 363-329.

[4] Kotelenez, P.: 1982, Law of large numbers and central limit theorem for chemical reactions with diffusions. Universität Bremen.

[5] Lenglart, E.: 1977, "Relations de domination entre deux processus", Ann. Inst. H. Poincaré, B XIII, 171-179.

[6] Métivier, M.: 1982, Semimartingales, De Gruyter, New York.

[7] Métivier, M.: 1984, "Convergence faible et principe d'invariance pour des martingales à valeurs dans des espaces de Sobolev", to appear in Ann. Inst. H. Poincaré.

[8] Métivier, M. and Joffe, A.: 1983, Weak convergence of sequences of semimartingales with applications to multitype branching processes, preprint.

[9] Rebolledo, R.: 1979, "La méthode des martingales appliquée à la convergence en loi des processus", Mémoires de la S.M.F., 62.

E. Pardoux

ASYMPTOTIC ANALYSIS OF A SEMI-LINEAR PDE WITH WIDE-BAND NOISE DISTURBANCES

ABSTRACT. We present an asymptotic analysis in the "white-noise limit" of a semi-linear parabolic partial differential equation, whose coefficients are perturbed by a wide-band noise. We prove the convergence in law towards the solution of an Ito stochastic PDE, thus generalizing the results in [3] for linear PDEs.

1. INTRODUCTION

We study here the "white noise limit" of a PDE, whose coefficients are perturbed by a wide band noise. We will obtain an Ito-type stochastic non linear PDE for the limit in law, as $\varepsilon \to 0$, of u^ε, the solution of :

$$\frac{\partial u^\varepsilon}{\partial t}(t,x)+Au^\varepsilon(t,x)+F(u^\varepsilon(t,x))+\frac{1}{\varepsilon}B(Z_{t/\varepsilon^2})u^\varepsilon(t,x)=$$

$$= f(x)+\frac{1}{\varepsilon}g(Z_{t/\varepsilon^2},x)$$

$t \geqslant 0$, $x \in D \subset \mathbb{R}^n$; $u^\varepsilon(t,x)=0$, $x \in \partial D$; $u^\varepsilon(0,x)=u_0(x)$

where $\forall z \in \mathbb{R}^d$, A and B(z) are linear partial differential operators, respectively of second and first order, $\{Z_t\}$ is a stationary positive recurrent diffusion process with values in $\mathbb{R}^d$ which perturbs the coefficients of B and g, with $E[B(Z_t)u(x)]= 0$, $E[g(Z_t,x)]=0$

for all u in some function space, and F is a Lipschitz

L. Arnold and P. Kotelenez (eds.), Stochastic Space-Time Models and Limit Theorems, 227–242.

mapping from $\mathbb{R}$ into $\mathbb{R}$.

Similar convergence results in the finite dimensional case - i.e. where (1.1) is replaced by an ODE - can be formd e.g. in PAPANICOLAOU [8], BLANKENSHIP-PAPANICOLAOU [2], KUSHNER [6]. This paper generalises the results of BOUC-PARDOUX [3], where $F \equiv 0$ (i.e. the PDE is linear). Convergence results for linear PDEs have also been obtained recently by KUSHNER and HUANG [7],[4].

Since the arguments are very similar to those in [3], we will only sketch here most of the proofs, giving some details concerning the treatment of the non-linear term F.

In section 2, we state our hypotheses concerning the diffusion process $\{Z_t\}$, and recall the results in [3] about its associated Poisson equation, which will be slightly generalized. In section 3, we formulate the precise assumptions on A, B, F, f, g, and prove the convergence of the law Q^ε of u^ε, towards the solution of a martingale problem.

2. THE RECURRENT DIFFUSION $\{Z_t\}$ AND ITS ASSOCIATED POISSON EQUATION.

2.1 The diffusion process $\{Z_t\}$.

$\{Z_t\}$ is supposed to be a diffusion process with values in $\mathbb{R}^d$, defined on a probability space (Ω_0, F_0, P), whose infinitesimal generator is given by :

$$L = \frac{1}{2}\,\overline{a}_{ij}(z)\frac{\partial^2}{\partial z_i \partial z_j} + \overline{b}_i(z)\,\frac{\partial}{\partial z_i}$$

Note that we use here and henceforth the convention of summation upon repeated indices. We now make some hypotheses on the coefficients $\overline{a}$ and $\overline{b}$:

(H1) $\overline{a}_{ij}$, $\overline{b}_i$, $\overline{a}_i \triangleq \overline{b}_i - \frac{1}{2}\frac{\partial \overline{a}_{ij}}{\partial z_j}$ are bounded C^∞ functions from $\mathbb{R}^d$ into $\mathbb{R}$; a_{ij} and b_i are uniformly Hölder continuous with some exponent $\alpha \in]0,1]$; $i,j = 1,\dots,d$.

(H2) $\exists \beta > 0$ s.t. $\overline{a}_{ij}(z)u_i u_j \geq \beta |u|^2$, $\forall z,\ u \in \mathbb{R}^d$

(H3) $\exists N > 0$ s.t. $\bar{b}_i(z)z_i|z|^{-1} \leqslant -N^{-1}$, $\forall z$ s.t. $|z| \geqslant N$

It follows from (H3)- see KHASMINSKII [5] - that the diffusion process $\{Z_t\}$ is positive recurrent. It then follows from (H1)-(H2) that its unique invariant probability measure possess a smooth density $p(x)$, s.t. $p(x) > 0$, $\forall x \in \mathbb{R}^d$. From now on, we suppose that this invariant measure is the law of Z_0.

Assume moreover :

(H4) $\exists N > 0$ s.t. $[|z|p(z)]^{-1} z_i \frac{\partial}{\partial z_j} [\bar{a}_{ij} p](z) \leqslant -N^{-1}$, $\forall z$ s.t. $|z| \geqslant N$

Remark 2.1 (H4) is equivelent to the more explicit condition (H3) whenever :

$$z_i[(2p(z))^{-1} \frac{\partial}{\partial z_j}(\bar{a}_{ij} p)(z) - \bar{b}_i(z)] \equiv 0$$

The latter equality holds in particular when the process $\{Z_t\}$ is "time-reversible", i.e. when $\forall t > 0$, $\forall \phi \in C_b(\mathbb{R}^d)$, $E[\phi(Z_t)/Z_0 = x] = E[\phi(Z_0)/Z_t = x]$, which is always the case if $d = 1$. □

Remark 2.2 We do not assume here, as we did in [3], that $p^{-1} a_{ij} \frac{\partial p}{\partial z_j}$ be bounded over $\mathbb{R}^d$, $\forall i = 1,\ldots,d$. This does not affect the proof of Theorem 2.3 below, but we will have to adapt the proof of Theorem 2.6. □

2.2 Solution of the associated Poisson equation.

Define $\hat{L}^2 \triangleq L^2(\mathbb{R}^d; p(z)dz)$, $\hat{L}^2_0 = \{u \in \hat{L}^2; \int p(z)u(z)dz = 0\}$, and $\hat{H}^1 \triangleq \{u \in \hat{L}^2; \frac{\partial u}{\partial z_i} \in \hat{L}^2, i=1,\ldots,d\}$. $(.,.)_\wedge$ and $|.|_\wedge$ will denote respectively the scalar product and norm of $\hat{L}^2$.

We consider L as an unbounded operator on $\hat{L}^2$, and denote $D(L) \triangleq \{u \in \hat{L}^2;\ Lu \in \hat{L}^2\}$.

Note that for $u \in D(L)$,

$$-(Lu, u)_\wedge = (\mathcal{L} u, u)_\wedge$$

where $\mathcal{L}$ is a self-adjoint non-negative operator on $\hat{L}^2$

defined by :

$$\mathcal{L} = - \frac{1}{2p} \frac{\partial}{\partial z_j} [\bar{a}_{ij}\, p \frac{\partial}{\partial z_i}(.)]$$

In fact, $\mathcal{L} = L$ if and only if $\{Z_t\}$ is time-reversible. From (H2), one can deduce :

$$(\mathcal{L}\, u,u)_\Lambda \geq \frac{\beta}{2} \sum_{i=1}^{d} \left|\frac{\partial u}{\partial z_i}\right|_\Lambda^2, \ \forall u \in D(L)$$

It then follows that the eigenspace associated with the eigenvalue 0 of $\mathcal{L}$ is the set of constants. We now quote several results from [3]:

<u>Theorem 2.3</u> Assume that (H1),(H2) and (H4) hold. Then o is an isolated point of the spectrum of $\mathcal{L}$. □

We note that, as mentionned in [3], the boundedness of the drift b is not necessary for Theorem 2.3 to hold.

<u>Corollary 2.4</u> Under the assumptions of Theorem 2.3, if $f \in \hat{L}^2_o$, then there exists a unique solution $v \in \hat{H}^1 \cap \hat{L}^2_o$ to the Poisson equation :

$$L v + f = o$$

Moreover, $v(z) = \int_o^\infty E_z[f(Z_t)]dt$, where E_z denotes expectation under the law of $\{Z_t\}$ conditionned upon $Z_o = z$. □

<u>Remark 2.5</u> It follows from Theorem 2.3 that there exists $\lambda > o$ s.t. $\forall f \in \hat{L}^2_o$ (i.e. orthogonal in $\hat{L}^2$ to the eigenspace associated to o), $|E.f(Z_t)|_\Lambda \leq e^{-\lambda t}|f|_\Lambda$, which insures the convergence of the integral $\int_o^\infty E_z[f(Z_t)]dt$. Note that the above inequality, with the $\hat{L}^2$-norm replaced by the sup-norm, would follow from Doeblin's condition, which is usually assumed for recurrent processes with values in a compact set, but can't be satisfied by a diffusion process with values in the whole space $\mathbb{R}^d$. □

We will need conditions on f which insure that v be bounded.

<u>Theorem 2.6</u> Assume (H1), (H2), (H3) and (H4).

Let $f \in \hat{L}^2_o \cap L^\infty(\mathbb{R}^d)$ satisfying the following condition :

$$\exists N > 0 \text{ s.t. } \int_N^{+\infty} \operatorname*{ess\,sup}_{|z| \geq t} |f(z)|\,dt < \infty$$

Then the solution $v \in \hat{L}^2_o \cap \hat{H}^1$ of $Lv + f = 0$ is a.e. bounded and $\exists\, c(N)$ s.t.

$$\| v \|_\infty \leq c(N)[\| f \|_\infty + \int_N^\infty \operatorname*{ess\,sup}_{|z| \geq t} |f(z)|\,dt] \quad (*)$$

where $\|.\|_\infty$ denotes the sup norm.

<u>Proof</u> : This is Theorem 1.12 of [3], which is due to P.L. LIONS. The proof given there uses the boundedness of $p^{-1}\bar{a}_{ij}\frac{\partial p}{\partial z_j}$. Now, from the smoothness and strict positivity of p, this quantity is still locally bounded. Therefore, instead of integrating by parts the quantity

$$(L\bar{w}, \bar{w}^-)_\wedge$$

we integrate by parts :

$$(L\bar{w}, \varphi\bar{w}^-)_\wedge$$

where φ is C^∞ with compact support, and ≥ 0.
We obtain :

$$0 \leq a_{B_o}(\bar{w}^-, \varphi\bar{w}^-) = -\frac{1}{2}(\bar{a}_{ij}\frac{\partial\overline{w}^-}{\partial z_i}, \varphi\frac{\partial\overline{w}^-}{\partial z_j})_\wedge -$$
$$-(\bar{a}_{ij}\frac{\partial\overline{w}^-}{\partial z_i}, \frac{\partial\varphi}{\partial z_j}\bar{w}^-)_\wedge - \frac{1}{4}(\bar{a}_{ij}\frac{\partial^2\varphi}{\partial z_i\partial z_j}\bar{w}^-, \bar{w}^-)_\wedge -$$
$$-\frac{1}{2}(\bar{a}_i\bar{w}^-, \frac{\partial\varphi}{\partial z_i}\bar{w}^-)_\wedge$$

It is then easy to take the limit as $\varphi \to 1$, giving the same conclusion as in [3]. Finally, the estimate (*) follows from a careful analysis of the proof. □

For $f \in \hat{L}^2_o$, we denote by Mf the element of $L^2_o \cap \hat{H}^1$ defined by:

$$(Mf)(z) = \int_o^\infty E_z f(Z_t)\,dt$$

i.e. $M = -L^{-1}$.

Lemma 2.7 Let $f \in \hat{L}^2_o(\mathbb{R}^d) \cap C^1_b(\mathbb{R}^d)$ [i.e. f is of C^1-class, and is bounded together with its first partial derivatives]. Suppose that $Mf \in L^\infty(\mathbb{R}^d)$.
Then $\frac{\partial}{\partial z_i} Mf \in L^\infty(\mathbb{R}^d)$, $i=1,\dots,d$; and $\exists c$ s.t.:

$$\left\| \frac{\partial Mf}{\partial z_i} \right\|_\infty \leqslant c \left[\left\| \frac{\partial f}{\partial z_i} \right\|_\infty + \| Mf \|_\infty \right] \quad \square$$

Corollary 2.8 There exists a constant c(N), which depends only on N and the operator L, s.t.

$$\left\| \frac{\partial Mf}{\partial z_i} \right\|_\infty \leqslant c(N) \left[\left\| \frac{\partial f}{\partial z_i} \right\|_\infty + \| f \|_\infty + \int_N^\infty \operatorname*{ess\,sup}_{|z| \geqslant t} |f(z)| dt \right]$$

$\square$

Let us make some comments in order to motivate what follows. Let first $f \in L^1(\mathbb{R}^d; p(z)dz)$. We have the following law of large numbers, which follows from a classical ergodic theorem :

$$\frac{1}{t} \int_o^t f(Z_s) ds \longrightarrow \int f(z)\, p(z) dz \quad \text{a.e.}$$

One might expect a central limit theorem of the following type to hold :
If $f \in \hat{L}^2_o$, $\frac{1}{\sqrt{t}} \int_o^t f(Z_s) ds \xrightarrow{\mathcal{L}} N(o, 2(f, Mf)_\wedge)$

where $Mf(z) = \int_o^\infty E_z f(Z_t) dt$. In fact, one might expect - see BHATTACHARYA [1], TOUATI [10] - a functional central limit theorem of the type :
If $f \in \hat{L}^2_o$, $\frac{1}{\varepsilon} \int_o^{\varepsilon^2 t} f(Z_s) ds \xrightarrow[\varepsilon \to o]{\mathcal{L}}$ Wiener process with

covariance coefficient $2(f,Mf)_\wedge$. In fact, a very particular case of the proof in section 3 establishes the above functional central limit theorem for f satisfying the conditions of Theorem 2.6.

Note that :

$$\frac{1}{\varepsilon} \int_o^{\varepsilon^2 t} f(Z_s) ds = \frac{1}{\varepsilon} \int_o^t f(Z_{s/\varepsilon^2}) ds$$

therefore, one might say, in a sense, that $\{\frac{1}{\varepsilon} f(Z_{t/\varepsilon^2})\}$

converges to a white noise, the time - derivative of a Wiener process. This formal statement may give some intuition for the convergence result of section 3.

3. CONVERGENCE OF THE PARABOLIC PDE WITH WIDE BAND NOISE DISTURBANCES.

3.1 Hypotheses and Notations

D will denote an open subset of $\mathbb{R}^n$, and $H_o^1(D)$ the closure of the set of C^∞ functions with compact support in D, with respect to the norm :

$$\| u \| = (|u|^2 + \sum_{i=1}^{n} | \frac{\partial u}{\partial x_i} |^2)^{1/2}$$

where $|.|$ denotes the usual norm in $L^2(D)$. Note that if the boundary ∂D of D is somewhat regular, $H_o^1(D)$ is exactly the subset of $H^1(D) \triangleq \{u \in L^2(D); \frac{\partial u}{\partial x_i} \in L^2(D);\ i=1,..,n\}$ consisting of those u which trace on ∂D is zero. We now identify $L^2(D)$ with its dual space, and the dual space of $H_o^1(D)$ with $H^{-1}(D)$. We denote by $< .,. >$ the pairing between $H_o^1(D)$ and $H^{-1}(D)$.

Suppose ave are given :

$$\text{(G.1)} \quad \begin{cases} A \in \mathcal{L}(H_o^1(D), H^{-1}(D)) \text{ s.t. } \exists \alpha > o\ ,\ \lambda \text{ with :} \\ <A u,u> + \lambda |u|^2 \geqslant \alpha \|u\|^2\ ,\quad \forall u \in H_o^1(D) \end{cases}$$

A could be $-\Delta = -\sum_{i=1}^{n} \frac{\partial^2}{\partial x_i^2}$, or more generally an elliptic second order operator.

For any $z \in \mathbb{R}^d$, $B(z)$ is a first order operator :

$$B(z) = b_i(x,z) \frac{\partial}{\partial z_i} + b_o(x,z)$$

(G.2) b_i is C^∞ in (x,z), the function and all its derivatives being bounded; $i=o,1,..,n$.

$$\text{(G.3)} \quad \begin{cases} \exists N > o \text{ s.t. } \int_N^\infty \operatorname{ess\,sup}_{|z| \geqslant t} (\sum_{|j| \leqslant d_i} | \frac{\partial^{|j|} b_i(x,z)}{\partial x_1^{j_1} \ldots \partial x_n^{j_n}} |) dt \leqslant N ; \\ i = o,1,..,n; \text{ with } d_o=2,\ d_i=3,\ i=1,..,n. \end{cases}$$

(G.4) $\int b_o(x,z)p(z)dz = \int b_i(x,z)p(z)dz = 0 \quad \forall x \in D,\ i=1,..,d$

(G.5) $\begin{cases} F: D \times \mathbb{R} \to \mathbb{R} \text{ is a measurable mapping satisfying :} \\ \text{(i)} \quad F(x,o) = o \\ \text{(ii)} \quad |F(x,u)-F(x,v)| \leqslant k\,|u-v| \end{cases}$

In the sequel, for $u \in L^2(D)$, we will denote by $F(u)$ the element of $L^2(D)$ defined by $F(u)(x)=F(x,u(x))$.

(G.6) $f \in L^2(D)$

(G.7) $\begin{cases} g: D \times \mathbb{R}^d \to \mathbb{R} \text{ is of class } C^1\text{, and the following} \\ \text{functions of } x \text{ are supposed to belong to } L^2(D): \\ \sup_z |g_i(.,z)|, \sup_z \left|\frac{\partial g_i}{\partial z_j}(.,z)\right|, \int_N^\infty \operatorname{ess\,sup}_{|z|\geqslant t} |g_i(.,z)|\,dt \\ \text{for } i=o,1,..,n;\ j=1...d;\ \text{where } g_o=g; g_i=\frac{\partial g}{\partial x_i}, i=1..n \end{cases}$

(G.8) $\int g(x,z)p(z)dz = 0, \quad \forall x \in D$

(G.9) $u_o \in L^2(D)$

For each $\varepsilon > 0$, we define $u_t^\varepsilon(\omega_o,x)$ as the unique solution of :

$$\left.\begin{aligned} \frac{du_t^\varepsilon}{dt} + A u_t^\varepsilon + F(u_t^\varepsilon) + \frac{1}{\varepsilon}B(Z_t^\varepsilon)u_t^\varepsilon &= f + \frac{1}{\varepsilon}g(Z_t^\varepsilon) \\ u_o^\varepsilon &= u_o \end{aligned}\right\} \quad (3.1)$$

which belongs P a.s. to $L^2(o,T;H_o^1(D)) \cap C([o,T];L^2(D))$, $\forall T > 0$. Here and in the sequel, $Z_t^\varepsilon \triangleq Z_{t/\varepsilon^2}$. Clearly, the operator and forcing terms in (3.1) could depend on t; we have avoided that in order to simplify the notations. In the next sections all hypotheses (H.1)-(H.4) and (G.1)-(G.9) are assumed to hold.

3.2 Weak compactness

From now on, we fix an arbitrary $T > 0$. $\overline{L^2(D)}$ will denote the space $L^2(D)$ endowed with its weak topology. We define :

$$\Omega = C([o,T];\ \overline{L^2(D)}) \cap L^2(o,T;H_o^1(D))$$

and endow Ω with the supremum of the topology of uniform convergence on $C([o,T]; L^2(D))$, and the weak topology on $L^2(o,T;H_0^1(D))$. Let F denote the Borel σ-field on Ω, and Q^ε the law $\{u_t^\varepsilon, t \in [o,T]\}$ on (Ω,F). We point out that also Ω is not a Polish space, Prohorov criterion for weak relative compactness of sets of probability measures on Ω holds-see VIOT [11].

On can prove-exactly as in [3]:

Theorem 3.1 The family of probability measures $\{Q^\varepsilon, \varepsilon > o\}$ on (Ω,F) is tight.

Sketch of the proof : We first note that a sufficient condition for $K \subset \Omega$ to be precompact is :

$$\sup_{u \in K} \sup_{t \in [o,T]} |u(t)| < \infty \qquad \text{(i)}$$

For each θ in a dense subset of $L^2(D)$, $\{t \to (u(t),\theta), u \in K\}$ is a set of equicontinuous mappings. (ii)

$$\sup_{u \in K} \int_o^T \| u(t) \|^2 dt < \infty \qquad \text{(iii)}$$

This indicates us that in order to prove the theorem, we need to show that $\exists c$ s.t:

$$E(\sup_{t \leqslant T} |u_t^\varepsilon|^2 + \int_o^T \|u_t^\varepsilon\|^2 dt) \leqslant c, \forall \varepsilon > o \qquad (3.2)$$

and also to take care of the modulus of continuity of $(u_t^\varepsilon, \theta)$, for θ in a dense subset of $L^2(D)$. The difficulty in establishing the estimate (3.2)comes from the $\frac{1}{\varepsilon}$ terms in the equation (3.1). The trick to kill them is to replace the quantity $|u_t^\varepsilon|^2$ to be estimated by :

$$|u_t^\varepsilon|^2 - 2\varepsilon(MB(Z_t^\varepsilon)u_t^\varepsilon, u_t^\varepsilon) + 2\varepsilon(Mg(Z_t^\varepsilon), u_t^\varepsilon)$$

where $M = - L^{-1}$ has been defined in §2. Applying Ito's formula to the above quantity, we obtain :

$$|u_t^\varepsilon|^2 - 2\,\varepsilon(MB(Z_t^\varepsilon)u_t^\varepsilon, u_t^\varepsilon) + 2\,\varepsilon(Mg(Z_t^\varepsilon), u_t^\varepsilon) +$$

$$+ 2\int_o^t < \tilde{A}(u_s^\varepsilon), u_s^\varepsilon > ds - 4\,\varepsilon \int_o^t < \tilde{A}(u_s^\varepsilon), M\tilde{B}(Z_s^\varepsilon)u_s^\varepsilon > ds +$$

$$+ 2\varepsilon\int_0^t < \tilde{A}(u_s^\varepsilon), Mg(Z_s^\varepsilon) > ds - 4\int_0^t (M\tilde{B}(Z_s^\varepsilon)u_s^\varepsilon, B(Z_s^\varepsilon)u_s^\varepsilon)ds +$$

$$+ 2\int_0^t (B(Z_s^\varepsilon)u_s^\varepsilon, Mg(Z_s^\varepsilon))ds = |u_0|^2 - 2\varepsilon(M\tilde{B}(Z_0)u_0, u_0) + 2\varepsilon(Mg(Z_0), u_0) +$$

$$+ 2\int_0^t (f, u_s^\varepsilon)ds + 2\varepsilon\int_0^t (Mg(Z_s^\varepsilon) - 2M\tilde{B}(Z_s^\varepsilon)u_s^\varepsilon, f)ds -$$

$$- 4\int_0^t (M\tilde{B}(Z_s^\varepsilon)u_s^\varepsilon, g(Z_s^\varepsilon))ds + 2\int_0^t (Mg(Z_s^\varepsilon), g(Z_s^\varepsilon))\, ds +$$

$$+ 2\int_0^t [(\nabla_z Mg(Z_s^\varepsilon), u_s^\varepsilon) - (\nabla_z MB(Z_s^\varepsilon)u_s^\varepsilon, u_s^\varepsilon)]\sigma(Z_s^\varepsilon)dW_s^\varepsilon$$

where $\tilde{A}(u) = Au + F(u)$, $\tilde{B} = \frac{1}{2}(B + B^*)$ is a zero order operator. (3.2) now follows by standard PDE arguments, using the hypotheses, in particular (G.1). The rest of the proof is identical to that in [3]. □

3.2 The limiting law

We now define the limiting law Q on (Ω, F).

Define $\overline{A} = A - \int_0^\infty E\,[B(Z_t)B(Z_0)]dt$

$$\overline{f} = f - \int_0^\infty E[B(Z_t)g(Z_0)]dt$$

Note that the differences between A and $\overline{A}$, f and $\overline{f}$ can be interpreted as Ito-Stratonovich correcting terms.
For $u \in H_0^1(D)$, let $R(u) \in \mathcal{L}(L^2(D))$ be given by :

$$(R(u)h,k) = \int_0^\infty E[(B(Z_0)u - g(Z_0), h)(B(Z_t)u - g(Z_t), k) +$$
$$+ (B(Z_t)u - g(Z_t), h)(B(Z_0)u - g(Z_0), k)]dt, \quad h,k \in L^2(D).$$

For $\omega \in \Omega$, we denote $v_t(\omega) = \omega(t)$. We will say that a probability measure Q on (Ω, F) solves the martingale problem $MP(u_0, \overline{A}, F, \overline{f}, R)$ if :

(i) $v_0(\omega) = u_0$, Q a.s.

(ii) $\forall\, \theta \in H_0^1(D)$, the process $\{M_t^\theta; t \in [o,T]\}$ given by :

$$M_t^\theta \triangleq (v_t, \theta) - (u_0, \theta) + \int_0^t < \overline{A}\, v_s + F(v_s) - \overline{f}, \theta > ds$$

is a continuous Q-martingale whose quadratic variation process is given by :

$$<M^\theta>_t = \int_0^t R(v_s)ds$$

By the same argument as in [3], one can show that $\exists K(u) \in \mathcal{L}(L^2(D))$ such that $u \to K(u)$ is linear, and $R(u) = K(u) \circ K(u)^*$.
It then follows that, if Q solves $MP(u_0, \bar{A}, F, \bar{f}, R)$, there exists a Q-cylindrical brownian motion $\{W_t, t \in [o,T]\}$ on $L^2(D)$ [$\forall h \in L^2(D)$, (W_t, h) is a Wiener process with associated increasing process $t|h|^2$], such that:

$$v_t + \int_0^t [\bar{A} v_s + F(v_s)]ds = u_0 + \int_0^t \bar{f}\, ds + \int_0^t K(v_s)dW_s \qquad (3.3)$$

From [9], (3.3) has a unique strong solution.
It then follows - see VIOT [11] - that the martingale problem $MP(u_0, \bar{A}, F, \bar{f}, R)$ is well-posed. In the sequel, Q will denote the unique solution of $MP(u_0, \bar{A}, F, \bar{f}, R)$.

Remark 3.2 In some cases, the martingale in the right-hand side of (3.3) can be written as an integral with respect to a finite-dimensional Wiener process. Let us consider the simplest example. Suppose that $B(z) = \alpha(z)B$ and $g(z) = \alpha(z)g$.
Then $(R(u)h, k) = 2(\int_0^\infty E[\alpha(Z_0)\alpha(Z_t)]dt)(Bu - g, h)(Bu - g, k)$.
Suppose (this is not a restriction) that $2\int_0^\infty E[\alpha(Z_0)\alpha(Z_t)]dt = 1$.
Then (3.3) reads :

$$v_t + \int_0^t (A - \frac{1}{2}B^2)v_s\, ds + \int_0^t F(v_s)ds = u_0 + \int_0^t (f - \frac{1}{2}Bg)ds + \int_0^t (Bv_s - g)dW_s$$

where here $\{W_t\}$ is a one dimensional standard Wiener process. □

3.3. Convergence of Q^ε

It remains to show that Q is the limit of any converging subsequence $\{Q^{\varepsilon_n}, n \in \mathbb{N}\}$, such that $\varepsilon_n \to o$. Here comes the difficulty with the non-linear term F. Indeed, the mapping :

$$v_. \quad \to \quad F(v_.)$$

is not continuous on Ω, since a continuous non linear mapping from $L^2(D)$ into itself is usually not weakly continuous. Our proof will follow the lines of that of VIOT [11, Thm II 1.1].

Define $\Omega' = L^2(]o,T[\times D)$, which we endow with its weak topology, $\overline{\Omega} = \Omega \times \Omega'$ and $\overline{F}$ the Borel σ-field on $\overline{\Omega}$. Let $\overline{Q}^\varepsilon$ denote the law of $\{u_t^\varepsilon, F(u_t^\varepsilon); t \in [o,T]\}$ on $(\overline{\Omega},\overline{F})$.
The tightness of $\{\overline{Q}^\varepsilon, \varepsilon > o\}$ follows immediately from that of $\{Q^\varepsilon, \varepsilon > o\}$. Define $r_t(\omega') = \omega'(t)$; By the argument of Theorem 2.8 in [3], we get that if $\overline{Q}$ is a weak-limit of a sequence $\{\overline{Q}^{\varepsilon_n}, n \in \mathbb{N}\}$, with $\varepsilon_n \to o$, then :

(i) $v_o(\omega) = u_o \quad \overline{Q}$ a.s.

(ii) $\forall \theta \in H_o^1(D)$, the process $\{\overline{M}_t^\theta(\omega,\omega'), t \in [o,T]\}$

given by :

$$\overline{M}_t^\theta \overset{\Delta}{=} (v_t,\theta) - (u_o,\theta) - \int_o^t \langle \overline{A}\, v_s + r_s - \overline{f}, \theta \rangle\, ds$$

is a continuous $\overline{Q}$-martingale whose quadratic variation process is given by :

$$\langle \overline{M}^\theta \rangle_t = \int_o^t R(v_s)\, ds$$

Clearly, since we are working on Lusin spaces, we have the desintegration :

$$Q(d\omega, d\omega') = S(d\omega)\, \mu_\omega\, (d\omega')$$

Suppose for a moment that we have established :

<u>Lemma 3.3</u> $\forall w \in L^2(o,T;H_o^1(D))$, S a.s.

$$\int_{\Omega'} \left(\int_o^T (r_t(\omega'), w(t))\, dt\right) \mu_\omega(d\omega') = \int_o^T (F(v_t(\omega)), w(t))\, dt \quad \square$$

It then follows that the process:

$$M_t^\theta(\omega) \overset{\Delta}{=} \int \overline{M}_t^\theta(\omega,\omega')\, \mu_\omega(d\omega')$$
$$= (v_t(\omega),\theta) - (u_o,\theta) - \int_o^t \langle \overline{A} v_s(\omega) + F(v_s(\omega)) - \overline{f}, \theta \rangle\, ds$$

is a continuous S-martingale, whose quadratic variation process is the same as that of $\overline{M}_t^\theta$. Then $S = Q$, the unique solution of $MP(u_o, \overline{A}, F, \overline{f}, R)$.
It is now easy to conclude :

Theorem 3.4 Let $\varepsilon_n \to 0$ be such that the sequence $\{Q^{\varepsilon_n}, n \in \mathbb{N}\}$ converges weakly. Then $Q^{\varepsilon_n} \Rightarrow Q$

Proof : We can extract a further subsequence $\{\varepsilon_m\}$, such that $\{\overline{Q}^{\varepsilon_m}, m \in \mathbb{N}\}$ converges weakly, as probability measures on $(\overline{\Omega}, \overline{F})$. From the above arguments,

$$\overline{Q}^{\varepsilon_m} \Rightarrow \overline{Q}$$

Then $$Q^{\varepsilon_m} = \int_{\Omega'} \overline{Q}^{\varepsilon_m}(.,d\omega') \Rightarrow \int_{\Omega'} \overline{Q}(.,d\omega') = Q$$

Proof of Lemma 3.3 $\overline{E}, \overline{E}^{\varepsilon}$ will denote respectively expectation with respect to $\overline{Q}, \overline{Q}^{\varepsilon}$. Let $\lambda > 0$. For $w \in \mathcal{V} = L^2(0,T;H_0^1(D))$, define :

$$X(\omega,\omega',w) \triangleq \int_0^T e^{-\lambda t}\{2<\overline{A}(v_t - w_t) + r_t - F(w_t), v_t - w_t> + \\ + \lambda\, |v_t - w_t|^2 - TrR(v_t - w_t)\}dt$$

Since $r_t(\omega') = F(v_t(\omega))$ $\overline{Q}^{\varepsilon}$ a.s., it follows from our standing hypotheses that we can choose λ large enough such that :

$$\overline{E}^{\varepsilon}(X) \geqslant 0, \forall w \in \mathcal{V} \qquad (3.4)$$

From now on, we fix λ such that (3.4) holds.

$$X = X_1 + X_2 \quad , \text{ where}$$

$$X_1 = \int_0^T e^{-\lambda t}\{2 <\overline{A}v_t + r_t, v_t> + \lambda\, |v_t|^2 - TrR(v_t)\}dt$$

Since $X_2 = X - X_1$ is continuous on $\overline{\Omega}$, $\forall w \in \mathcal{V}$, and $\overline{E}^{\varepsilon}(X_2^2) \leqslant c$ $\forall \varepsilon > 0$, it is easy to show that, provided $\overline{Q}^{\varepsilon_n} \Rightarrow \overline{Q}$,

$$\overline{E}^{\varepsilon_n}(X_2) \to \overline{E}(X_2) \qquad (3.5)$$

We now look for another expression for $\overline{E}^{\varepsilon}(X_1)$.
A slight modification of the computation made in the proof of Theorem 3.1 leads to :

$$e^{-\lambda T}|u_T^{\varepsilon}|^2 + 2\int_0^T e^{-\lambda t} < \tilde{A}(u_t^{\varepsilon}) - MB(Z_t^{\varepsilon})B(Z_t^{\varepsilon})u_t^{\varepsilon}, u_t^{\varepsilon} > dt + \\ + \lambda \int_0^T e^{-\lambda t}|u_t^{\varepsilon}|^2 dt = |u_0|^2 + 2\int_0^T e^{-\lambda t}(f - MB(Z_t^{\varepsilon})g(Z_t^{\varepsilon}), u_t^{\varepsilon})dt + \\ + 2\int_0^T e^{-\lambda t}(MB(Z_t^{\varepsilon})u_t^{\varepsilon} - Mg(Z_t^{\varepsilon}), B(Z_t^{\varepsilon})u_t^{\varepsilon} - g(Z_t^{\varepsilon}))\, dt + Y_{\varepsilon}$$

where $E(Y_\varepsilon)=0(\varepsilon)$. Let $\Phi(z,u)$ be the solution of the Poisson equation :

$$L\,\Phi(z,u)=E(MB(Z_0)u-Mg(Z_0)u,B(Z_0)u-g(Z_0))-$$
$$-(MB(z)u-Mg(z),B(z)u-g(z))+$$
$$+E\langle MB(Z_o)B(Z_o)u,u\rangle-\langle MB(z)B(z)u,u\rangle-$$
$$-E(MB(Z_0)g(Z_0),u)+(MB(z)g(z),u)$$

Apply Ito formula to the process $2\varepsilon^2 e^{-\lambda t}\Phi(Z_t^\varepsilon,u_t^\varepsilon)$, integrate from o to T, and add to the above identity. Taking the expectation in the resulting identity yields :

$$\overline{E}^\varepsilon(X_1)=|u_0|^2+2\,\overline{E}^\varepsilon\int_0^T e^{-\lambda t}(\overline{f},v_t)dt-e^{-\lambda T}\overline{E}^\varepsilon(|v_T|^2)+0(\varepsilon)$$

Since the mapping $v\to|v_T|^2$ is lower semi-continuous on Ω, we conclude that :

$$\overline{\lim_{n\to\infty}}\;\overline{E}^{\varepsilon n}(X_1)\leqslant|u_0|^2+2\,\overline{E}\int_0^T e^{-\lambda t}(\overline{f},v_t)dt-e^{-\lambda T}\overline{E}(|v_T|^2)\quad(3.6)$$

On easily deduce from the caracterization of Q that :

$$\overline{E}(X_1)=|u_0|^2+2\,\overline{E}\int_0^T e^{-\lambda t}(\overline{f},v_t)dt-e^{-\lambda T}\overline{E}(|v_T|^2)\qquad(3.7)$$

It now follows froms (3.4),(3.5),(3.6) and (3.7) :

$$\overline{E}(X)\geqslant 0$$

Let m be any Borel probability measure on $\mathcal{V}$, Φ any non negative bounded continuous functional on $\mathcal{V}$. We have :

$$\int_V \Phi(w)\int_{\Omega} X(\omega,\omega',w)\overline{Q}(d\omega\,,\,d\omega')m(dw)\geqslant o$$

Using Fubini's theorem and the change of variable $w=v(\omega)-\rho\,\overline{w}$, $\rho>o$, we deduce :

$$\int_{V\times\overline{\Omega}}\Phi(v-\rho\overline{w})\frac{1}{\rho}X(\omega,\omega',v(\omega)-\rho\,\overline{w})m_\omega(d\overline{w})\overline{Q}(d\omega,d\omega')\geqslant o$$

where $m_\omega(d\overline{w})$ is a certain transition probability. We can then let $\rho\to o$ by Lebesgue's dominated convergence theorem, yielding :

$$\int_{V\times\overline{\Omega}}\Phi(v(\omega))\Big(\int_0^T e^{-\lambda t}(r_t(\omega')-F(v_t(\omega)),\overline{w}_t)dt\Big)m_\omega(d\overline{w})\overline{Q}(d\omega,d\omega')$$

The result follows from the freedom of choice for Φ and m. □

CONCLUSION : We have used here the "monotonicity method" for proving a convergence result for a non linear PDE with wide-band noise input. One should be able to handle other types of nonlinearities by means of a compactness argument (see VIOT [11]), but we don't know what is really possible in that direction. In any case, non linear terms of the type $\frac{1}{\varepsilon} G(Z_t^\varepsilon, u_t^\varepsilon)$ in equation (3.2) seen to be significantly more difficult to treat than what we have done here.

E. Pardoux
UER de Mathématiques
Université de Provence
3, Pl. V. Hugo, 13331 Marseille Cedex 3
and Laboratoire de Mécanique et d'Acoustique

REFERENCES

[1] R.N. BHATTACHARYA 'On the Functional Central Limit Theorem and the law of the Iterated Logarithm for Markov Processes' *Z. Wahrscheinlichkeitstheorie* 60, 185 - 201 (1982).

[2] G. BLANKENSHIP, G. PAPANICOLAOU 'Stability and Control of Stochastic Systems with Wide-Band Noise Disturbances I'. *SIAM J. Appl. Math.* 34, 437 - 476 (1978).

[3] R. BOUC, E. PARDOUX 'Asymptotic Analysis of PDEs with Wide-Band Noise Disturbances, and Expansion of the Moments' *Stoch. Anal. and Applic.*, to appear.

[4] H. HUANG, H. KUSHNER 'Weak Convergence and Approximations for Partial Differential Equations with Stochastic Coefficients. Preprint.

[5] R. KHASMINSKII *Stochastic Stability of Differential Equations*, Sijthoff and Noordhoff (1980).

[6] H. KUSHNER 'Junp-Diffusion Approximations for Ordinary Differential Equations with Wide-Band Random Right Hand Side' *SIAM J. Control and Opt.* 17, 729-744 (1979) .

[7] H. KUSHNER, H. HUANG 'Limits for Parabolic Partial Differential Equations with Wide-Band Stochastic Coefficients' *Stochastics*, to appear.

[8] G. PAPANICOLAOU 'Asymptotic Analysis of Stochastic Equations' in *Studies in Probability Theory*, M. Rosenblatt Ed., MAA Studies in Math. 18, 111-179 (1978).

[9] E. PARDOUX 'Equations aux Dérivées Partielles Stochastiques Nonlinéaires Monotones' Thesis, Univ. Paris XI (1975).

[10] A. TOUATI 'Théorèmes de Limite Centrale Fonctionnels pour les Processus de Markov' *Ann. Inst. Henri Poincaré*, 19, 43 - 55 (1983).

[11] M. VIOT 'Solutions Faibles d'Equations aux Dérivées Partielles Stochastiques Non linéaires', Thesis, Univ. Paris VI (1976).

Hermann Rost

A CENTRAL LIMIT THEOREM FOR A SYSTEM OF INTERACTING PARTICLES

1. INTRODUCTION

Consider a system of particles moving at random in $\mathbb{Z}^d$ in such a way that its probability law is stationary in time and invariant under spatial shifts; denote by $X(j,t)$ the number of particles at site j at time t, let ρ be its expectation. The corresponding <u>fluctuation processes</u> are the $\mathcal{S}'$-valued processes N^ε, defined by

$$(1)\qquad N^\varepsilon(\phi,t)=\varepsilon^{d/2}\cdot\sum_j \phi(\varepsilon j)(X(j,t\varepsilon^{-2})-\rho),\qquad \phi\in\mathcal{S}$$

($\mathcal{S}'$: space of Schwartz distributions on R^d,
$\mathcal{S}$: space of rapidly decaying smooth functions).

The parameter ε ranges between 0 and 1; it measures the ratio of microscopic to macroscopic length unit. In [1] for a particular class of models it has been shown that

$$(2)\qquad \lim_{\varepsilon\to 0} EN^\varepsilon(\phi,0)N^\varepsilon(\psi,t)=\chi\cdot\int\phi\cdot T_t\psi\cdot dx\quad ,$$

where (T_t) is the semigroup on R^d generated by $\kappa/2\cdot\Delta$, and χ and κ are explicitly calculable in terms of the given dynamics and the equilibrium density ρ.

As is well known, there exists a Gaussian process N, indexed by R_+ and $\mathcal{S}$, whose covariance is given by the r.h.s. of (2). It may also be characterized by the following two properties:

(3a) $\quad EN(\phi,t)^2=\chi\int\phi^2\cdot dx,\qquad \phi\in\mathcal{S}\ ,\ t\geq 0\ ;$

(3b) $\quad$ for each $\phi\in\mathcal{S}$, the process $t\to N(\phi,t)-N(\phi,0)-\int_0^t\kappa/2\cdot N(\Delta\phi,s)ds$ is a Brownian motion with diffusion constant $-u\int\phi\cdot\Delta\phi\cdot dx$ where $u=\chi\cdot\kappa/2$

(see e.g. [2],[3]); we call it the stationary <u>Ornstein-Uhlenbeck process</u> with drift operator $\kappa/2\cdot\Delta$ and <u>driving noise operator</u> $-u\cdot\Delta$.

We show in this short note that in addition to convergence

L. Arnold and P. Kotelenez (eds.), Stochastic Space-Time Models and Limit Theorems, 243–248.

of the covariances, as stated in [1], the laws of the processes N^{ε}, as measures in the Skorokhod space $\mathcal{D}(R_+,\mathcal{S}')$, converge weakly to that of the process N characterized by (3). We show even more, namely that this convergence already takes place in the space $\mathcal{D}(R_+,\mathcal{H}(-d-4))$. (For notations, see section 2.) In particular, we thus establish the limiting joint Gaussian character of the variables $N^{\varepsilon}(\phi,t)$, $t\geq 0$, $\phi\in\mathcal{S}$. A similar result, for independent particles, has been proven in [2]; it is natural that our proof has many arguments in common with that paper.

2. ASSUMPTIONS AND NOTATIONS

The underlying particle process X is Markovian on a subset of the space of counting measures on Z^d; its generator is

$$(4)\qquad Lf(x)=\sum_j c(x_j)\cdot\sum_{|k-j|=1}(f(x+\delta_k-\delta_j)-f(x)).$$

The dynamics is determined by the function c which is defined on the positive integers and is supposed to be increasing and Lipschitz. (For existence, see e.g. [4].) We consider the process X only at a fixed equilibrium with distribution at a single time μ_u, where μ_u is the product measure with identical factors, satisfying

$$(5)\qquad \mu_u(x_j=m)=Z(u)^{-1}\cdot u^m\cdot\prod_1^m c(n)^{-1},\quad m\in N .$$

Here u is a positive parameter characterizing the equilibrium measure. The following expectations are of interest to us

$$(6)\qquad \rho:=\mathcal{E}X(j,t)$$

$$\chi:=\mathcal{E}(X(j,t)-\rho)^2\ ;$$

we compute, besides

$$(7)\qquad u=\mathcal{E}c(X(j,t))\ .$$

In order to fomulate the concept of weak convergence we need the following notations (as in [3]): $\mathcal{H}(m)$, $m\geq 0$, is the completion of the space of infinitely differentiable functions on R^d with compact support w.r. to the norm

$$(8)\qquad \|\phi\|_m:=(\int\phi\cdot(x^2-\Delta)^m\phi\cdot dx)^{1/2}\ .$$

The dual of $\mathcal{H}(m)$ will be denoted by $\mathcal{H}(-m)$. We have the relations

$$\mathcal{S}=\bigcap_m \mathcal{H}(m)\,, \qquad \mathcal{S}'=\bigcup_m \mathcal{H}(-m)\ .$$

We use the symbol ν for elements of the spaces $\mathcal{H}(-m)$, the value of ν at $\phi\in\mathcal{H}(m)$ is written as $\nu(\phi)$; random elements in $\mathcal{H}(-m)$ are denoted by N, N^{ε} etc. If we introduce the system of Hermite functions h_α, $\alpha=(\alpha_1,\ldots,\alpha_d)$ a multiindex, each element ϕ of $\mathcal{H}(m)$ allows an expansion

(9) $$\phi=\sum_\alpha c(\alpha)h_\alpha \quad \text{with } \sum c(\alpha)^2\cdot(2|\alpha|+d)^m<\infty$$

This expansion is meaningful for negative m, too, if for m<O and $\nu\in\mathcal{H}(m)$ we interprete the statement

$$\nu=\sum_\alpha c(\alpha)h_\alpha \quad \text{by} \quad c(\alpha)=\nu(h_\alpha) \qquad \text{for all } \alpha.$$

The fluctuation processes N^{ε} have already been defined in (1). Here we remark only that we can interprete each N^{ε} even as $\mathcal{H}(-m)$-valued process for sufficiently large m (in section 3 we take m=d+4); correspondingly, the evaluation $N^{\varepsilon}(\phi,t)$ in (1) is meaningful for $\phi\in\mathcal{H}(m)$, not only $\phi\in\mathcal{S}$.

3. THE CONVERGENCE THEOREM

In this section we prove the following

Theorem. The laws of the processes N^{ε} converge weakly, as $\varepsilon\to 0$, in the Skorokhod space $\mathcal{D}(R_+,\mathcal{H}(-d-4))$ to the law of the Ornstein-Uhlenbeck process N defined by (3).

Proof. We begin with an outline of the major steps. We represent, for each ϕ, $N^{\varepsilon}(\phi,\cdot)$ as semimartingale and estimate its drift and martingale part (A). From there, we conclude that up to any finite time T with high probability the processes N^{ε} can be uniformly approximated in $\mathcal{H}(-d-4)$-norm by a process with values in a finite dimensional space (B). Hence tightness of the family N^{ε} is reduced to tightness of $N^{\varepsilon}(\phi,\cdot)$ for each ϕ separately (C). Finally, using the result established in [1], we show that any limiting process satisfies the conditions (3), which proves the desired convergence statement.

(A) From the definition of the dynamics it follows that for each ϕ the processes $M_i^{\varepsilon}(\phi,\cdot)$, i=1,2, are martingales:

(10a) $M_1^\varepsilon(\phi,t):=N^\varepsilon(\phi,t)-N^\varepsilon(\phi,0)-\int_0^t A^\varepsilon(\phi,s)ds,$

where

(10b) $A^\varepsilon(\phi,t)=\varepsilon^{d/2}\sum_j\left[c(X(j,t\varepsilon^{-2}))-u\right]\cdot\Delta^\varepsilon\phi(\varepsilon j)$

$(\Delta^\varepsilon\phi(y):=\sum_{|e|=1}\varepsilon^{-2}\cdot(\phi(y+\varepsilon\cdot e)-\phi(y)));$

(11a) $M_2^\varepsilon(\phi,t):=(M_1^\varepsilon(\phi,t))^2-\int_0^t B^\varepsilon(\phi,s)ds$,

where

(11b) $B^\varepsilon(\phi,t):=\varepsilon^d\cdot\sum_j c(X(j,t\varepsilon^{-2}))\cdot\sum_{|k-j|=1}\varepsilon^{-2}\cdot(\phi(\varepsilon k)-\phi(\varepsilon j))^2;$

further , the jump size of $M_1^\varepsilon(\phi,\cdot)$ is bounded by

$$\varepsilon^{d/2+1}\sup_y|\nabla\phi(y)| .$$

We have the estimate for T fixed:

(12a) $E\sup_{t\leq T}(\int_0^t A^\varepsilon(\phi,s)ds)^2$

$\leq E(\int_0^T|A^\varepsilon(\phi,s)|ds)^2\leq T^2\cdot E(A^\varepsilon(\phi,0))^2$,

by stationarity. An elementary calculation shows that the right hand side is bounded by

(12b) $T^2\cdot\text{var }c(X(0,0))\cdot\left\{\|\phi\|_2^2+\varepsilon^2\cdot K_1\|\phi\|_3^2\right\}$

with K_1 depending only on d. The variance of c(X(0,0)) is finite, as follows easily from (5) and the Lipschitz property of c(.). Similarly

(13) $E(M_1^\varepsilon(\phi,T))^2=T\cdot EB^\varepsilon(\phi,0)=$

$\leq T\cdot Ec\cdot(\|\phi\|_1^2+K_2\cdot\varepsilon^2\cdot\|\phi\|_2^2)$.

B. From the preceding estimate follows the existence of a constant K(T) such that, uniformly in $0<\varepsilon<1$,

(14) $E\sup_{t\leq T}(N^\varepsilon(\phi,t))^2\leq K(T)\cdot\|\phi\|_3^2$.

We choose ϕ equal to the Hermite function h_α and get

$$(15)\quad E \sup_{t\leq T} (N^\varepsilon(h_\alpha,t))^2 \leq K(T)\cdot(2|\alpha|+d)^3$$

If we multiply (15) by $(2|\alpha|+d)^{-d-4}$ and sum over α, the r.h.s. converges; hence, in the norm of $H(-d-4)$, the processes N^ε are, uniformly in ε, approximated by their projections on a finite dimensional subspace.

C. In order to show tightness for fixed $\phi\in S$ of the family $N^\varepsilon(\phi,\cdot)$, we do it separately for $M_1^\varepsilon(\phi,\cdot)$ and $\int A^\varepsilon(\phi)ds$. The second family is tight, because as in (12a)

$$(16)\quad E(\int_r^t A^\varepsilon(\phi,s)ds)^2 \leq (t-r)^2\cdot K(\phi)\ ,$$

which implies tightness even in $C(R_+,R)$ by the Kolmogorov criterion.
The martingales $M_1^\varepsilon(\phi)$ are convergent to Brownian motion, hence tight; indeed, $B^\varepsilon(\phi,s)$ converges in L^2 to $u\cdot\|\phi\|_1^2$, therefore the integrals $\int_0^t B^\varepsilon(\phi,s)ds$ converge to $t\cdot u\|\phi\|_1^2$. Since the maximal jump size of $M_1^\varepsilon(\phi)$ tends to zero (A), the claim follows ([5]).

D. By th. 3 in [1], we have

$$(17)\quad \lim_{\varepsilon\to 0} \int_0^t (A^\varepsilon(\phi,s)-u/\chi\cdot N^\varepsilon(\Delta\phi,s))ds = 0$$

in L^2. (In this argument the specific properties of the model have been used; it does not follow from pure Sobolev space reasoning.) Hence any limit process N of a sequence $N_{\varepsilon'}$, $\varepsilon'\to 0$, must satisfy, because of (10), (11), and (C),

$$(18)\quad N(\phi,t)-N(\phi,0)-\int_0^t u/\chi N(\Delta\phi,s)ds$$

is Brownian motion with diffusion constant $u\cdot\|\phi\|_1^2$

Since obviously the relation

$$(19)\quad \lim_{\varepsilon\to 0} EN^\varepsilon(\phi,t)^2 = \chi\cdot\int\phi^2\cdot dy$$

holds, any limit process N is identified as the stationary Ornstein-Uhlenbeck process described in (3). This proves the theorem.

H. Rost
Institut für Ang. Math.
Im Neuenheimer Feld 294
6900 Heidelberg
West Germany

REFERENCES

1 Th. Brox, H. Rost: 'Equilibrium fluctuations of stochastic particle systems: the role of conserved quantities.' Ann.Probability 12 (1984),

2 A. Martin-Löf: 'Limit theorems for the motion of a Poisson system of independent Markovian particles at high density.' Z.Wahrscheinlichkeitstheorie verw. Geb. 34 (1976), 205-223.

3 R. Holley, D. Stroock: 'Generalized Ornstein-Uhlenbeck processes and infinite particle Brownian motion.' Publ. RIMS Kyoto Univ. 14 (1978), 741-788.

4 Th. Liggett: 'An infinite particle system with zero range interaction.' Ann.Probability 1 (1973), 240-253.

5 R.S. Liptser, A.N. Shiryayev: 'A functional central limit theorem for semimartingales.' Th. Probability Appl. 25 (1980), 667-688.

Hans Zessin

MOMENTS OF STATES OVER NUCLEAR LSF SPACES

Introduction: The purpose of the present paper is to investigate the problem of weak convergence of a sequence of states over nuclear LSF spaces V (including $\mathcal{S}(\mathbb{R}^\nu)$ and $\mathcal{D}(\mathbb{R}^\nu)$) from assumptions on the asymptotic behaviour of their moments; here by a state over V we understand a continuous linear random function over V. This paper is closely related to [16] where the same problem has been studied for random measures; it is also closely related to the papers of Fernique [4] as well as Dobrushin/Minlos [3], where the basic tools on linear random functions are developed.

The main results, a uniqueness and continuity theorem for the moments of states, can be found in § 1: In theorem 1.1 Carleman's one-dimensional uniqueness condition (1.2) assures uniqueness of a linear process Z in terms of its moments $(\nu_Z^{2n})_n$. This condition is: $\sum_{n\geq 1} \nu_Z^{2n}(f,\overset{2n}{\ldots},f)^{-1/2n}$ diverges for each $f \in V$. The continuity theorem 1.3 establishes weak convergence of a sequence of states (P_n) to some limiting state under the assumptions that for each k the k-th moments of P_n converge to some limit, that these limit moments satisfy Carleman's more-dimensional uniqueness criterium (1.8), and finally under condition (1.9) resp. (1.9') that yields relative compactness of $\{P_n\}$. Theorem 1.2 gives a partial converse of theorem 1.3.

In § 2 we consider the special case $V = \mathcal{S}(\mathbb{R}^\nu)$ and give a sufficient condition for relative compactness of $\{P_n\}$ in terms of the second moments, namely: $\sup_n \nu_{P_n}^2(f,f) \leq C\cdot\|f\|_m^2$ for each $f \in \mathcal{S}(\mathbb{R}^\nu)$ for some $C > 0$ and $\mathcal{S}(\mathbb{R}^\nu)$ - norm $\|\cdot\|_m$. We then comment on the results obtained in connection with the existence problem of the φ_ν^4 - model of Euclidean quantum field theory.

Acknowledgement: I am grateful to M. Röckner for useful discussions.

L. Arnold and P. Kotelenez (eds.), Stochastic Space-Time Models and Limit Theorems, 249–261.

§ 1 A Uniqueness and Continuity Theorem for Moments of States over nuclear LSF spaces

Let V be a linear topological space and V^* its dual equipped with weak topology. V^*_{alg} denotes the algebraic dual of V. For $f \in V$, $\mu \in V^*_{alg}$ define $\xi_f(\mu) = \mu(f)$. Let $\mathcal{B}_0(V^*_{alg}) = \sigma(\xi_f ; f \in V)$ be the smallest σ-algebra of subsets in V^*_{alg} with respect to which all the functions ξ_f, $f \in V$, are measurable. For $f \in V$, $\mu \in V^*$ we set $\zeta_f(\mu) = \mu(f)$ and denote by $\mathcal{B}_0(V^*) = \sigma(\zeta_f ; f \in V)$ the σ-algebra in V^* generated by the functions ζ_f, $f \in V$.

Definition: A linear map Z from V into the set of random variables over a probability space $(\Omega,\mathcal{F},\mathcal{P})$ is called a linear process over V (on $(\Omega,\mathcal{F},\mathcal{P})$). Given a probability measure P on $(V^*,\mathcal{B}_0(V^*))$ the linear process Z over V on $(V^*,\mathcal{B}_0(V^*),P)$ defined by $Z : f \to \zeta_f$ is called a state over V (given by P). We also call a probability measure P on $(V^*,\mathcal{B}_0(V^*))$ a state over V. A sequence (P_n) of states over V converges weakly to a state P, and we write $P_n \Rightarrow P$, if $P_n(\varphi) \to P(\varphi)$ for each bounded continuous real function φ on V^*. Two linear processes Z_1,Z_2 over V on probability spaces $(\Omega_1,\mathcal{F}_1,\mathcal{P}_1)$, $(\Omega_2,\mathcal{F}_2,\mathcal{P}_2)$ are said to be (probabilistically) equivalent, if for all $k \in \mathbb{N}$, $f_1,\ldots,f_k \in V$ the corresponding k-dimensional distributions coincide, i.e.

$$\mathcal{P}_{1,(Z_1(f_1),\ldots,Z_1(f_k))} = \mathcal{P}_{2,(Z_2(f_1),\ldots,Z_2(f_k))} .$$

A linear process Z over V on $(\Omega,\mathcal{F},\mathcal{P})$ is called of k-th order, if for each $f_1,\ldots,f_k \in V$

$$\nu_Z^k(f_1,\ldots,f_k) = \int_\Omega Z(f_1) \cdot \ldots \cdot Z(f_k) \; d\mathcal{P} \tag{1.1}$$

exists and is finite. In this case ν_Z^k is called the moment of Z of k-th order. For a state P over V we write also ν_P^k. Z is called of infinite order if it has moments of all orders.

By Hölder's inequality a sufficient condition for a linear process Z over V to be of k-th order is
$\int_\Omega |Z_f|^k \, d\mathcal{P} < +\infty$ for each $f \in V$.

Theorem 1.1: Let Z be a linear process over V on $(\Omega,\mathcal{F},\mathcal{P})$ of infinite order. Suppose that

$$(1.2) \qquad \sum_{n=1}^{\infty} \nu_Z^{2n} (f,\overset{2n}{\ldots},f)^{-1/2n} = +\infty \quad \text{for each } f \in V ,$$

then Z is uniquely determined by its moments up to equivalence.

Proof: It is well known that for two linear processes over V to be equivalent it is necessary and sufficient that their one-dimensional distributions coincide. But (1.2) just means that the one-dimensional distributions of Z satisfy Carleman's criterium (see [16], Lemma 1.4) and therefore are uniquely determined by its moments.

Example 1.1: One important example of a linear process over V is the Gaussian process. This is a linear process Z over V s.th. for each $f \in V$ the random variable $Z(f)$ is Gaussian. It is well known that its moments are

$$(1.3) \qquad \nu_Z^k (f_1,\ldots,f_k) = \sum_{\mathcal{J}=\{J_1,\ldots,J_\ell\}} \prod_{i=1}^{\ell} \mathcal{P}\Big(\prod_{j\in J_i} Z(f_j)\Big) ,$$

where $\mathcal{J}$ is summed over all pair partitions of $\{1,\ldots,k\}$. By a pair partition we mean a partition $\{J_1,\ldots,J_\ell\}$ of $\{1,\ldots,k\}$ into disjoint subsets J_i, s.th., in the case when k is even, each J_i has two elements and in the case when k is odd, exactly one J_i has a single element. The rest has two elements.

$$(1.4) \qquad \nu_Z^{2n} (f,\overset{2n}{\ldots},f) = \frac{(2n)!}{2^n \cdot n!} \cdot \nu_Z^2 (f,f)^n , \quad f \in V ,$$

which immediately implies that condition (1.2) is satisfied, so that by theorem 1.1 Z is uniquely determined (up to equivalence) by its moments and thus, on account of (1.3), by its first and second moment.

Theorem 1.2: Let (P_n) be a sequence of states over V of infinite order. If P_n converges weakly to some state P and if

$$(1.5) \qquad \sup_n \nu_{P_n}^k (f,\overset{k}{\ldots},f) < +\infty , \quad k \in \mathbb{N} , \ f \in V .$$

then P is of infinite order and $\nu^k_{P_n}(f_1,\ldots,f_k) \underset{n}{\to} \nu^k_P(f_1,\ldots,f_k)$, $k \in \mathbb{N}$, $f_1,\ldots,f_k \in V$.

The proof is left to the reader.

In the following we assume that V is nuclear and an <u>LSF-space</u>, i.e. the strict inductive limit of a sequence of separable Frechét spaces. Note that the space $\mathcal{S}(\mathbb{R}^\nu)$ is a nuclear separable Frechét space; the space $\mathcal{D}(\mathbb{R}^\nu)$ is a nuclear LSF space (see [13]). Recall that in a nuclear space V the topology can be generated by a family $(q_\alpha)_{\alpha\in A}$ of seminorms, each of which originates from a positive semi-definite Hermitian form on $V \times V$ (see [13]). Recall also that $\mathcal{B}_0(V^*)$ coincides with the σ-algebra $\mathcal{B}(V^*)$ of Borel subsets of V^* if V is an LSF space (see [3], Prop. 1, e.g.).

We denote by $(q^*_\alpha)_{\alpha\in A}$ the family of seminorms on V^* defined by

$$(1.6) \qquad q^*_\alpha(\mu) = \sup\{|\mu(f)| : q_\alpha(f) \leq 1\} \ , \quad \mu \in V^* \ .$$

<u>Theorem 1.3:</u> Let V be a nuclear LSF space and let (P_n) be a sequence of states over V , each of which is of infinite order. Suppose that

(1.7) for each $k \in \mathbb{N}$, $f_1,\ldots,f_k \in V$ the limits $\nu^k(f_1,\ldots,f_k) = \lim_{n\to\infty} \nu^k_{P_n}(f_1,\ldots,f_k)$ exist and

$$(1.8) \qquad \sum_{n=1}^{\infty} \Big[\sum_{i=1}^{k} \nu^{2n}(f_i,\overset{2n}{\ldots},f_i)\Big]^{-1/2n} = +\infty \ , \quad k \in \mathbb{N} \ , \quad f_1,\ldots,f_k \in V \ .$$

If finally

(1.9) there exists $\alpha \in A$ s.th.

$\lim_{N\to\infty} \sup_n P_n\{q^*_\alpha > N\} = 0$ or equivalently

(1.9') for each $\varepsilon > 0$ and $a > 0$ there exists a neighborhood U of 0 in V s.th.

$$\sup_n P_n \{\sup_{f\in U}|\zeta_f| > a\} < \varepsilon \quad ,$$

then there exists a unique state P over V having moments of all orders s.th. $P_n \Rightarrow P$ and $\nu_P^k = \nu^k$ for each $k \in \mathbb{N}$. Moreover P is k-continuous, i.e. the mapping $f \to \zeta_f$ is continuous with respect to the topologies in V and $L_k(P)$, $k \in \mathbb{N}$.

Proof: 1. We first prove the equivalence of (1.9) and (1.9').
(1.9) $\Rightarrow$ (1.9'): Let $a,\varepsilon,\delta > 0$ and set $U_{\alpha,\delta} = \{f \in V : q_\alpha(f) < \delta\}$. Then $P_n \{\sup_{f\in U_{\alpha,\delta}} |\zeta_f| > a\} \leq P_n \{q_\alpha^* > \frac{a}{\delta}\}$.
Choosing δ in such a way that $\sup_n P_n \{q_\alpha^* > \frac{a}{\delta}\} < \varepsilon$, yields (1.9').
(1.9') $\Rightarrow$ (1.9): For $a,\varepsilon > 0$ choose a neighborhood U of 0 in V s.th. $\sup_n P_n \{\sup_{f\in U}|\zeta_f| > a\} < \varepsilon$. Now there exists α and N s.th. $\frac{a}{N} \{q_\alpha \leq 1\} \subseteq U$. Then
$\sup_n P_n \{q_\alpha^* > N\} \leq \sup_n P_n \{\sup_{f\in U}|\zeta_f| > a\} < \varepsilon$.
2. Let $k \in \mathbb{N}$, $r_1,\dots,r_k \in \mathbb{N}_0$, $f_1,\dots,f_k \in V$. Combining assumption (1.7) with lemma 1.4 of [16] yields the existence of a probability measure $\mu_{f_1,\dots,f_k}$ on $\mathbb{R}^k$ which is a solution of the ($\mathbb{R}^k$) moment problem corresponding to the moments

$$\alpha_{f_1,\dots,f_k}(r_1,\dots,r_k) = \nu^{r_1+\dots+r_k}(\overbrace{f_1,\dots,f_1}^{r_1},\dots,\overbrace{f_k,\dots,f_k}^{r_k}),$$

because

$$\sum_{r_1=0}^{N_1} \dots \sum_{r_k=0}^{N_k} a_{r_1,\dots,r_k} \cdot \alpha_{f_1,\dots,f_k}(r_1,\dots,r_k) =$$

$$= \lim_{n\to\infty} \int_{\mathbb{R}^k} \sum_{r_1=0}^{N_1} \dots \sum_{r_k=0}^{N_k} a_{r_1,\dots,r_k} \cdot x_1^{r_1}\dots x_k^{r_k}\, P_{n,(\zeta_{f_1},\dots,\zeta_{f_k})}(dx)$$

is nonnegative if the polynomial in the last integral is nonnegative on $\mathbb{R}^k$. Thus

$$(1.10) \qquad \nu^{r_1+\ldots+r_k}(f_1,\overset{r_1}{\ldots},f_1,\ldots,f_k,\overset{r_k}{\ldots},f_k) = \\ = \int_{\mathbb{R}^k} x_1^{r_1}\ldots x_k^{r_k}\ \mu_{f_1,\ldots,f_k}(dx)$$

for each $r_1,\ldots,r_k \in \mathbb{N}_0$.

3. From Carleman's more-dimensional uniqueness criterium (see lemma 1.4 of [16], e.g.) we know that the probability measures $\mu_{f_1,\ldots,f_\alpha}$ are uniquely determined by ist moments on account of assumption

$$(1.8) \qquad \sum_{n\geq 1} \Big(\sum_{i=1}^{k} \nu^{2n}(f_i,\overset{2n}{\ldots},f_i)\Big)^{-1/2n} = +\infty \ .$$

Therefore

$$(1.11) \qquad P_{n,(\zeta_{f_1},\ldots,\zeta_{f_k})} \Rightarrow \mu_{f_1,\ldots,f_k}\ , \quad f_1,\ldots,f_k \in V\ ,$$

follows from (1.7) combined with (1.10) by lemma 1.5 in [16].

4. The next step is to construct a probability measure P on $\mathcal{B}_0(V^*_{alg})$ having the $\mu_{f_1,\ldots,f_k}$ as marginal distributions: Let F be a k-dimensional subspace of V and $f_1,\ldots,f_k$ a basis of F . Define a probability measure P_F on $(V^*_{alg},(\zeta_{f_1},\ldots,\zeta_{f_k})^{-1}(\mathcal{B}_k))$ by

$$(1.12) \qquad P_F((\zeta_{f_1},\ldots,\zeta_{f_k})^{-1}(A)) = \mu_{f_1,\ldots,f_k}(A)\ , \quad A \in \mathcal{B}_k\ .$$

Here $\mathcal{B}_k$ denotes the σ-algebra of Borel subsets in $\mathbb{R}^k$. P_F is well defined, since $(\zeta_{f_1},\ldots,\zeta_{f_k})$ is surjective. Moreover (1.11) immediately implies that the family $\{P_F$; F finite-dimensional subspace of $V\}$ is consistent. Therefore we can define an additive probability measure P on the Boolean algebra $\widetilde{\mathcal{B}} = \cup_F\ \sigma(\xi_f\,;\, f\in F)$ in V^*_{alg} by

$$(1.13) \qquad P(B) = P_F(B) \quad , \quad B \in \sigma(\xi_f \, ; f \in F) \quad .$$

Now it is well known that P is σ - additive on $\tilde{\mathcal{B}}$. (Lenard's theorem; see [12] e.g.) Therefore $\tilde{P}$ can be extended to a probability measure P on $\sigma(\tilde{\mathcal{B}}) = \mathcal{B}_0(V^*_{alg})$. We note also that the linear process Z over V on $(V^*_{alg}, \mathcal{B}_0(V^*_{alg}), P)$ given by $Z : f \to \xi_f$ has moments of all orders s.th. $\nu_Z^k = \nu^k$, $k \in \mathbb{N}$. This follows immediately from (1.10).

5. We now show that P in continuous, i.e. $f_\gamma \to f$ in V implies $\xi_{f_\gamma} \to \xi_f$ in measure, i.e. $P\{|\xi_{f_\gamma}| > a\} \to 0$ for each $a > 0$. Since for each $f \in V$, $P_{n,\zeta_f} \Rightarrow P_{\xi_f}$ we have for each $a > 0$ and $f \in V$

$$P\{|\xi_f| > a\} \le \liminf_{n\to\infty} P_n \, \{|\zeta_f| > a\}$$
$$\le \sup_n P_n \, \{\sup_{f \in \mathcal{U}} |\zeta_f| > a\}$$

for each neighborhood $\mathcal{U}$ of 0 in V containing f . Continuity of P thus follows from (1.9'). Therefore by Minlos' theorem the nuclearity of V implies that Z is equivalent to a state over V which we again denote by P (see [3] or [4]).

6. Finally we show that $P_n \Rightarrow P$. As we know that

$$P_{n,(\zeta_{f_1},\dots,\zeta_{f_k})} \Rightarrow P_{(\zeta_{f_1},\dots,\zeta_{f_k})} \;, \; (k \in \mathbb{N} \;, \; f_1,\dots,f_k \in V),$$

it suffices to show that $\{P_n\}$ is relatively compact with respect to weak topology. For this it is sufficient that $\{P_n\}$ is tight, i.e. for each $\varepsilon > 0$ there is a compact subset K of V^* s.th. $P_n(K) \ge 1 - \varepsilon$ for each n . This follows from Prohorov's theorem which is valid in our situation, since V^* is the weak dual of an LSF space and thus a regular, standard Borel space (see [4] or [14]). We construct K in the following way: By the Banach-Steinhaus theorem a subset $X \subset V^*$ is weakly precompact if and only if it is equicontinuous, i.e. given $\varepsilon > 0$ there exists a neighborhood $\mathcal{U}$ of 0 in V s.th. $\sup_{f \in \mathcal{U}} \sup_{\mu \in X} |\mu(f)| \leq \varepsilon$. But by assumption (1.9') for $\varepsilon > 0$, $k \in \mathbb{N}$ there exists a neighborhood $\mathcal{U}_k$ of 0 in V s.th. for each n $P_n \{\sup_{f \in \mathcal{U}_k} |\zeta_f| > \frac{1}{k}\} \leq \frac{\varepsilon}{2^k}$. Thus

$P_n(\bigcap_{k \geq 1} \{ \sup_{f \in \mathcal{U}_k} |\zeta_f| \leq \frac{1}{k}\}) \geq 1 - \varepsilon$ for each n . Set

$K_\varepsilon = \bigcap_{k \geq 1} \{ \sup_{f \in \mathcal{U}_k} |\zeta_f| \leq \frac{1}{k}\}$. K_ε is equicontinuous and therefore weakly precompact. Thus $K = \overline{K}_\varepsilon$ is compact in V^* and has the required properties.

7. Continuity for each k follows at once from prop. 4 in [3] because P is of infinite order. This completes the proof of the theorem.

We now discuss the problem of weak convergence to a Gaussian process P .

Corollary 1.1: Let (P_n) be a sequence of states over a nuclear LSF space V of infinite order satisfying (1.7) and (1.9) resp. (1.9'). If furthermore the limiting moments ν^n , $n \in \mathbb{N}$, appearing in (1.7) obey (1.3), then there exists a unique Gaussian state P over V with mean value ν^1 and sec.moment ν^2 s.th. $P_n \Rightarrow P$. Moreover, P is k-continuous for each $k \in \mathbb{N}$.

Proof: 1. Note that $(\nu^{2n})_n$ satisfies the uniqueness condition (1.8) on account of (1.4). Thus we are in the situation of theorem 1.3. That the limiting state P is Gaussian with mean ν^1 and sec.moment ν^2 follows immediately from theorem 1.1 .

§ 2 The case $V = \mathcal{S}(\mathbb{R}^\nu)$, $\nu \geq 1$

We first summarize some facts about Schwartz distributions: (cf. [8] and [9]): The inner product and the norm in $L_2(\mathbb{R}^\nu)$ are denoted by (,) and $\| \ \|$ respectively. Given $n \in \mathbb{N}_0 = \mathbb{N} \cup \{0\}$ the Hermite function of order n corresponding to the Hermite polynomial H_n of order n is given by

(2.1) $$h_n(x) = c_n \cdot H_n \cdot \exp(-\frac{x^2}{2}) \ , \ x \in \mathbb{R} \ ,$$

where c_n is a constant for which $\|h_n\| = 1$. It is well known that $\{h_n \, ; n \geq 0\}$ is an orthonormal base in $L_2(\mathbb{R})$. Define for $\beta = (\beta_1,\ldots,\beta_\nu) \in \mathbb{N}_0^\nu$ the function $h_\beta : \mathbb{R}^\nu \to \mathbb{R}^\nu$ by

(2.2) $$h_\beta(x_1,\ldots,x_\nu) = h_{\beta_1}(x_1) \cdot \ldots \cdot h_{\beta_\nu}(x_\nu) \ .$$

$\{h_\beta \, ; \beta \in \mathbb{N}_0\}$ is an orthonormal base in $L_2(\mathbb{R}^\nu)$. The m-norm $\| \ \|_m$ is $\mathcal{S}(\mathbb{R}^\nu)$ is defined as follows:

(2.3) $$\|f\|_m^2 = \sum_\beta (f,h_\beta)^2 \cdot (2|\beta| + \nu)^m \ , \ f \in \mathcal{S}(\mathbb{R}^\nu) \ , \ m \in \mathbb{N}_0 \ ,$$

where $|\beta| = \beta_1 + \ldots + \beta_\nu$. $\mathcal{S}(\mathbb{R}^\nu)$ is topologized by the set of norms $\| \ \|_m$, $m \in \mathbb{N}_0$, and is nuclear as such. It is a real pre-Hilbert space with an inner product corresponding to $\| \ \|_m$. The (-m) - norm $\| \ \|_{-m}$ in $\mathcal{S}^*(\mathbb{R}^\nu)$ is defined as follows:

(2.4) $$\|\mu\|_{-m}^2 = \sum_\beta \mu(h_\beta)^2 \frac{1}{(2|\beta| + \nu)^m} \ , \ \mu \in \mathcal{S}^*(\mathbb{R}^\nu) \ .$$

It is easy to see that for each $m \in \mathbb{N}_0$, $\mu \in \mathcal{S}^*(\mathbb{R}^\nu)$

(2.5) $$\|\mu\|_{-m} = \sup \{ |\mu(f)| : f \in \mathcal{S}(\mathbb{R}^\nu) \ , \ \|f\|_m \leq 1 \} \ .$$

The definition of $\| \ \|_{-m}$ suggests a sufficient condition for (1.9) in terms of the second moments. Indeed we have

Theorem 2.1: Let (P_n) be a sequence of states over $\mathcal{S}(\mathbb{R}^\nu)$, $\nu \geq 1$. If

(2.6) for each $k \in \mathbb{N}$ there exists $C > 0$ and $m \in \mathbb{N}_0$ s.th.

$$\sup_n \nu_{P_n}^k (f,\overset{k}{\ldots},f) \leq C \cdot \|f\|_m^k \ , \ f \in \mathcal{S}(\mathbb{R}^\nu) \ ,$$

then there exists a state P over $\mathcal{S}(\mathbb{R}^\nu)$ of infinite order and a subsequence $(P_{n_\ell})_\ell$ s.th. $P_{n_\ell} \Rightarrow P$ and

$$\nu^k_{P_{n_\ell}}(f_1,\ldots,f_k) \underset{\ell}{\to} \nu^k_P(f_1,\ldots,f_k) \ , \ k \in \mathbb{N} \ , \ f_1,\ldots,f_k \in \mathcal{S}(\mathbb{R}^\nu).$$

Conversely, let each P_n be of infinite order satisfying (1.7) and (1.8). If furthermore

(2.7) there exists $C > 0$ and $m \in \mathbb{N}_0$ s.th.

$$\sup_n \nu^2_{P_n}(f,f) \leq C \cdot \|f\|^2_m \ , \ f \in \mathcal{S}(\mathbb{R}^\nu) \ ,$$

then there exists a unique state P over $\mathcal{S}(\mathbb{R}^\nu)$, k-continuous for each k , of infinite order s.th. $P_n \Rightarrow P$ and $\nu^k_P = \nu^k$, $k \in \mathbb{N}$.

Proof: 1. We first treat the second part of the theorem. It is easy to see that there exists $m' \geq m$ s.th.

$$\Sigma_\beta \|f_\beta\|^2_m < +\infty \ , \text{ where } \ f_\beta = \frac{h_\beta}{(2|\beta| + \nu)^{m'/2}} \ .$$

Therefore on account of (2.7)

$$\sum_\beta \sup_n \nu^2_{P_n}(f_\beta,f_\beta) < +\infty \ .$$

But this in turn implies (1.9) because

$$\sup_n P_n \{\| \ \|_{-m'} > N\} \leq \frac{1}{N^2} \sup_n P_n(\| \ \|^2_{-m'})$$

$$\leq \frac{1}{N^2} \cdot \sum_\beta \sup_n \nu^2_{P_n}(f_\beta,f_\beta) \underset{N\to\infty}{\to} 0 \ .$$

Thus theorem 1.3 applies.

2. Note that (2.6) implies (2.7) and therefore via (1.9) relative compactness of $\{P_n\}$. Thus there exists a state P over $\mathcal{S}(\mathbb{R}^\nu)$ and a subsequence (P_{n_ℓ}) s.th. $P_{n_\ell} \Rightarrow P$. Now theorem 1.2 applies and proves the first part.

Combining the second part of theorem 2.1 with Corollary 1.1 we have

Corollary 2.1: Let (P_n) be a sequence of states over $\mathcal{S}(\mathbb{R}^\nu)$, $\nu \geq 1$, of infinite order satisfying (1.7) and (2.7). If furthermore the limiting moments ν^n , $n \in \mathbb{N}$, appearing in (1.7) obey (1.3) then there exists a unique Gaussian state P over $S(\mathbb{R}^\nu)$ with mean value ν^1 and sec.moment ν^2 s.th. $P_n \Rightarrow P$. Moreover, P is k-continuous for each $k \in \mathbb{N}$.

Remarks: (1) Condition (2.7) is fullfilled for the socalled φ_ν^4 - lattice field models P_n of Euclidean quantum field theory. The limit of the sequence (P_n) considered there is the socalled continuum resp. scaling limit. For the details we refer to [2], [11] and [15]. The main problem of Euclidean quantum field theory is twofold: (α) to establish the existence of the continuum resp. scaling limit P of the sequence (P_n) and (β) to show that P is non-trivial, i.e. non-Gaussian. Here the second part of the problem is the much more difficult and is not answered yet for dimension $\nu = 4$.

The following remarks should show that theorem 2.1 combined with recent results of Palmer/Tracy [11] give a complete answer to the existence problem for dimension $\nu = 2$ (in the case of Ising-models instead of φ_2^4 - models), and, combined with a result of Glimm/Jaffe [7]; a not completely satisfactory answer in the other cases.

(2) For a sequence of two-dimensional Ising-models (P_n) Palmer/Tracy [11] verified conditions (1.7) and (1.8). Since (2.7) is also satisfied, by the second part of theorem 2.1 the scaling limit P of (P_n) exists and is uniquely determined. This limit is non-Gaussian.

(3) Consider a sequence (P_n) of φ_ν^4 - models, $\nu \geq 1$. We have remarked that (2.7) is true. Now by socalled Gaussian domination (see Newman [10]) condition (2.7) implies bounds for the moments $\nu_{P_n}^k(f,\ldots,f)$, $k \in \mathbb{N}$, $f \in S(\mathbb{R}^\nu)$, uniformly in n, in such a way that (2.6) holds. This has been remarked by Glimm/Jaffe [7]. Thus the first part of theorem 2.1 applies. To summarize: In a sequence (P_n) of φ_ν^4 - lattice field models with arbitrary $\nu \geq 1$ one can find a sequence $(P_{n_\ell})_\ell$ which in the scaling limit converges weakly to some state P over $S(\mathbb{R}^\nu)$ of infinite order s.th. the socalled Schwinger functions $\nu_{P_{n_\ell}}^k(f_1,\ldots,f_k)$ converge to the corresponding Schwinger function of P. This improves the result of Glimm and Jaffe. For dimension $\nu = 2,3$ it has been shown that P is non-Gaussian. (For a recent elementary proof of this fact see [6]). As has

been remarked by Fröhlich et al. [6] this result is non-satisfactory: "Aside from its inherent non-constructiveness, certain natural and desirable properties (such as uniqueness of the limit) go unestablished."

(4) We finally remark that Aizenman [1] and Fröhlich [5] have shown that a sequence of φ_ν^4 lattice field models satisfies the assumptions of Corollary 2.1 if $\nu > 4$. Thus in this case the weak scaling resp. continuum limit P is Gaussian.

Hans Zessin
Universität Bielefeld
Fakultät für Mathematik
Postfach 86 40
4800 Bielefeld 1
Federal Republik of Germany

References:

[1] Aizenman, M.: 1982, 'Geometric analysis of φ^4 fields and Ising models', Commun. math. Phys. 86, 1-48.

[2] Brydges, D.: 1982, 'Field theories and Symanzik's polymer representation', in: Gauge theories: fundamental interactions and rigoros results. Dita, P., Georgescu, V., Purice, R., eds. Boston-Basel-Stuttgart: Birkhäuser.

[3] Dobrushin, R.L., Minlos, R.A.: 'The moments and polynomials of a generalized random field'. Theory of probability and its applications 23, (1978), 686-699.

[4] Fernique, X.: 1967, 'Processes linéaires, processes généralisés'. Ann. Inst. Fourier 17, 1-92.

[5] Fröhlich, J.: 1982, 'On the triviality of $\lambda\phi_d^4$ theories and the approach to the critical point in $d \underset{(=)}{\geq} 4$ dimensions'. Nucl. Phys. B200, 281-296.

[6] Fröhlich, J., Brydges, D.C., Sokal, A.D.: 1983, 'A new proof of the existence and nontriviality of the continuum ϕ_2^4 and ϕ_3^4 quantum field theories'. Preprint.

[7] Glimm, J., Jaffe, A.: 1974, 'Remark on the existence of φ_4^4'. Phys. review letters 33, 440-442.

[8] Holley, R.A., Stroock, D.W.: 1978, 'Generalized Ornstein-Uhlenbeck processes and infinite particle branching Brownian motions'. Publ. RIMS, Kyoto Univ., 14, 741-788.

[9] Ito, K.: 1983, 'Distribution-valued processes arising from independent Brownian motions'. Math. Z. 182, 17-33.

[10] Newman, C.M.: 1975, 'Gaussian correlation inequalities for ferromagnets'. Z. Wahrscheinlichkeitstheorie verw. Gebiete 33, 75-93.

[11] Palmer, J., Tracey, C.: 1981, 'Two-dimensional Ising correlations: Convergence of the scaling limit'. Adv. appl. math. 2, 329-388.

[12] Reed, M.C.: 1973, 'Functional analysis and probability theory'. In: Constructive quantum field theory. Velo. G., Wightman, A. eds.. Berlin-Heidelberg-New-York: Springer.

[13] Schäfer, H.H.: 1971, 'Topological vector spaces'. New York: Springer.

[14] Smolyanov, O.G., Fomin, S.V.: 1976, 'Measures on topological linear spaces', Uspekki Matem. Nauk 31, 3-56.

[15] Sokal, A.D.: 1982, 'An alternate constructive approach to the φ_3^4 quantum field theory, and a possible destructive approach to φ_4^4'. Ann. Inst. Henn Poincaré A 37, 317-398.

[16] Zessin, H.: 1983, 'The method of moments for random measures'. Z. Wahrscheinlichkeitstheorie verw. Gebiete 62, 395-409.

Subject Index